MISSING MOUNTAINS

Missing Mountains

We Went to the Mountaintop But It Wasn't There

Introduction by Silas House

Afterword by Wendell Berry

Edited by Kristin Johannsen,
Bobbie Ann Mason,
and Mary Ann Taylor-Hall

WIND PUBLICATIONS

First edition

International Standard Book Number 1893239497
Library of Congress Control Number 2005933867

.

Front cover photographs copyright © Geoff Oliver Bugbee
www.geoffbugbee.com

The editors thank the Carnegie Center for Literacy and Learning in Lexington for making its space available for their meetings.

They shall not hurt nor destroy
in all my holy mountain. . . .
—Isaiah 11

Photographs in this book are from the following sources:

Warren Brunner — pages iii, 4, 10, 19, 23, 34, 46, 58, 62, 64, 73, 77, 82, 85, 95, 107, 108, 116, 131, 167, 193, 200, 202

Geoff Oliver Bugbee — front cover photos

Kentuckians For The Commonwealth — pages 138, 142, 147

Ann Olson — pages 13, 15, 28, 50, 90, 111, 129, 135, 140, 150, 158, 188, 195

Contents

Part Three

Preface

This book deals with a subject of the gravest importance: the destruction of the earth, in particular, the destruction of a large part of Kentucky. Kentucky's mountains and the creatures who live there are being devastated by the coal-mining technique known as mountaintop removal.

Think of this book as a scrapbook: a collection of essays, poems, stories, and personal testimonials by Kentuckians who oppose the terrible violence being inflicted on our home planet and native state.

Introduction

by Silas House

Coal mining is a part of me. My grandfather lost his leg in the Leslie County deep mines in the 1940s. When he was able, he went back into the mines and worked twenty more years.

My Uncle Sam was in a mining accident that left him branded by a coal tattoo across his left cheekbone. I can recall other uncles, cousins, and brothers-in-law coming home from the mines with the coal dust so thick in their lashes that it looked as if they had applied mascara. They all loved their jobs. Mining allowed them to rise out of poverty.

I am proud of my grandfather's lost leg. Proud of my uncle's coal tattoo. They are symbols of determination and hard work.

But slowly, my love for the industry turned to a love-hate relationship. When I was a teenager, there was a strip mine directly across from our house. We breathed the dust and listened to the groan of machinery for more than a year. I spent long hours on the ridge above the mines, watching and mourning the loss of the woods and rolling pasture I had played in all my life.

It wouldn't have been so bad if the land had been treated respectfully. But it wasn't. Trees were thrown aside like useless things. The good topsoil was buried beneath clay and rock. Still, I knew that coal mining was an important part of our economy.

A couple of years ago, I got my first glimpse of mountaintop removal—in which the summit of a mountain is removed to extract coal—along Kentucky Highway 80 in Knott County. The mountain that just the spring before had been crowded with a thousand redbuds was now a barren plateau dotted by shoots of brown grass and struggling saplings.

In April of 2005, a group of writers met on Lower Bad Creek to view and learn about active mountaintop removal. These were writers who are widely known and those who are just starting out, all Kentuckians, and all concerned citizens.

We walked through a healthy forest near the mining and viewed the wealth of herbs, plants, trees and water that was being threatened on all sides. We looked down on a strip mine. We drove through the valley and saw plateaus that had once been mountains on either side of us.

We drove twenty miles to Hindman. I counted eight mountains that had been removed along the road. Gone forever. Some were still dusty, noisy messes of bulldozers and exposed coal seams. Others had been reclaimed, but I saw no evidence of healthy forests or fertile pastures there.

At a town meeting in Hindman, we were greeted by a standing-room-only crowd of people who had come to share the stories of their experiences with mountaintop removal. These people live with mountaintop removal every day; they are a part of the land. It is their stories that matter.

There were tales of water that ran red as blood with sulfur. A man had drilled five wells over the last year because the mining blasts caused every one of them to go dry. A young woman lives on a road so damaged by coal trucks that ambulances aren't able to reach the older people who live there. Several people told of reporting damage to government officials, only to be told that the flooding, damaged foundations, and polluted air were all "acts of God."

None of these people were there with a vendetta against the coal industry. They were there because they wanted their stories taken to a larger audience. They were there because they care about their children and their grandchildren and because the land is a part of them, too.

It is mind-boggling that the whole nation is talking about Alaska being drilled for oil, yet no one cares that Appalachia has been systematically scalped for the last 28 years. And the speed with which that is happening is increasing daily. As one woman at the town meeting said, "I don't care what anybody says, the Arctic Circle isn't a bit more worthy of respect than my mountains."

We are not against the coal industry. Coal was mined for decades without completely devastating the entire region. My family is a part of that coal-mining legacy. But mountaintop removal means that fewer and fewer people work in mining, because it is so heavily mechanized. If mountaintop removal is banned, there might actually be more mining jobs for the hard-working people of Kentucky. And beyond that, the proper respect might finally be returned to the spirit of the land and its people.

You hold in your hands an emotional book. These are not emotional reactions to the issue of mountaintop removal. These are not emotional tirades. But these are editorials, poems, stories, and essays written with passion, compassion, fear, hope, and love—all emotions that go to the heart of this matter. These are pieces written by people who believe in helping others, in preserving the environment, and in making our world a better place.

Of all these emotions, the one that these pieces bring to mind in particular is hope.

When I first witnessed mountaintop removal up close, I was filled with a sense of dread, defeat, and remorse. I mourned the loss of the mountains, the trees, and the spirit of the people who had been damaged in the onslaught of this wasteful and disrespectful form of coal mining. The sight made me physically ill. Despite how sickened I was by the spectacle of Appalachia being systematically raped, I also immediately felt a glimmer of hope rising up, hope that a group of concerned citizens could get together and do something about the act of mountaintop removal.

Thousands of tired, nerve-shaken, over-civilized people are beginning to find out that going to the mountain is going home; that wildness is necessity; that mountain parks and reservations are useful not only as fountains of timber and irrigating rivers, but as fountains of life.

—John Muir

That is the goal of this book, to make the general public aware of the problem, and to build hope for all the people who suffer in the shadow of mountaintop removal. Although our primary focus is the people and land of Eastern Kentucky, these writers are also concerned about the effect of this form of mining on the rest of the state, the nation, the world. This practice chips away at the very heart of our planet.

Unfortunately, not enough people have witnessed mountaintop removal with their own eyes, and photographs simply do not convey its scope. With the rapid spread of mountaintop removal, many more people will be taking notice as the sites come closer to the highways, closer to the towns and rivers, closer to the public eye. But by then, it will be too late for too many mountains, too many streams, too many people.

It is important to point out that this is not only a book about the loss of a place, but also about the suffering of people. Readers often look at the word "environment" and automatically think of streams, trees, mountains. But an environment is also made up of people. This is a collection of writing that remembers the children who do not have good water to drink or bathe in, the people who travel unsafe roads or live beneath sites that have already sent boulders crashing through their homes. This book calls to account a government that prefers to produce coal for our energy-consuming nation in the quickest, cheapest way, rather than to find a safe, more efficient and respectful method, one which would also create jobs for the region. These are writers who support private land owners, independent truckers, and working people, writers who are hopeful of finding a better, less wasteful, and more respectful way of creating a progressive and viable economy.

We trust that this book will help to raise awareness of what is happening in our own back yards or headed our way in one form or another—polluted air or water or the deadened spirit of one of the most beautiful places on earth.

Contained herein are the emotional words of respect and reverence for a place and a way of life that is going to disappear if we don't do something now. This writing is often sad, always important.

These are words of hope that we might all come together and stop the disrespect to our land. These are words that stand on their own and tell you what you need to know about the crisis confronting our people and our land in the face of mountaintop removal.

I shall not leave these prisoning hills . . .
Being of these hills, being one with the fox
Stealing into the shadows, one with the new-born foal,
The lumbering ox drawing green beech logs to mill,
One with the destined feet of man climbing and descending,
And one with death rising to bloom again, I cannot go.
Being of these hills I cannot pass beyond.
—from "Heritage" by James Still

PART ONE

Appalachia Extinct

by Lucy Flood

On April 20th, 2005, my father and I left our house just as the sun topped the walnut trees at our farm's eastern edge. We were headed for the mountaintop removal tour organized by Kentuckians For The Commonwealth (KFTC). Our group of sixteen Kentucky writers would spend two days observing how coal companies and legislators had despoiled an entire region of the country.

Looking southeast from our farm, you can see the foothills of the Appalachian Mountains etched in jagged blue silhouettes against the horizon. Nearly 320,000 acres of Kentucky have been strip-mined in the last thirty years. That morning as I gazed out at the mountains, I was aware that the people of Appalachia have paid dearly with their health and quality of life to live in their hauntingly beautiful mountains, where coal runs in seams. I wanted to hear the testimonies of people who live every day in the shadow of mountaintop removal.

At our meeting spot, I settled into a van with three other writers, a photographer, and Jerry Hardt, the KFTC trip organizer. As we headed east to Leslie County, I looked all around. The mountains had erupted in the stunning color of redbuds. In one spot, I saw a barn so overgrown with vines that I couldn't tell whether any of the original structure remained or whether the vines had simply formed a skeleton in the shape of a vanished barn. Jerry pointed out spots where only a thin corridor of pristine mountains still lies between the highways and the moonscapes that mountaintop removal operations leave behind. Like the barn's vines, the row of hills masked reality: the coal companies have systematically torn down the mountains that once filled all of Eastern Kentucky.

Just past Hyden, we crossed over Greasy Creek. At the top of a hill, dogwoods and redbuds cast the forest in a dazzling pink-and-white pattern. The wind blew, sending white blossoms tripping down out of the dogwoods. We hit the gravel, and the van jolted. A truck with "Flammable" painted down its side tore by us, sending stones clinking against our vehicle.

We stopped just over a culvert. To our right, a barn and a house sat in a valley where a freshly tilled field exposed its rich earth. Mining had crippled a mountain at the property's far edge, lopping off its peaks. To our left, we saw a silt pond created by a mining company that had long since gone

bankrupt and then restarted under another name, to avoid clean-up costs. I learned that coal companies often use this trick to avoid paying for the true cost of mining.

Our host, Daymon Morgan, emerged up the driveway. After returning from World War II, he'd bought this land on Lower Bad Creek in Leslie County for $1,000. I saw pride in the tilt of his straw cowboy hat, in the shine on his black cowboy boots, and in the way he stood up straight, smiling, as he looked out at the valley. Mr. Morgan climbed into his six-wheeled ATV and drove a group of us up a gravel road, while the others followed on foot.

When we could drive no further, we got out and walked down the wet path. As we converged around Mr. Morgan, he said, "It's very disturbing for me to see the things I love being destroyed. I got my medicine and food from these mountains, and I still do. There's a place down here where I can lay down and drink out of the creek and I want to keep it that way because it's clear up above. I feel like I'm being pushed into a corner."

Mr. Morgan bent over, running his fingers through the forest's dark soil. He explained to us that it takes tens of thousands of years for the forest's organic matter to biodegrade into such rich earth. Straightening up, he said, "Just right across the mountain there's mining. We're fighting terrorism, but what about taking care of our people right here at home? The U.S. has the cheapest electric rates of anywhere. But look what it's costin' us. We don't need money to be coming in that way."

10

He led us through the forest, pointing out medicinal plants and showing why biologically diverse forests should be protected. As Mr. Morgan rolled a ginger root between his fingers, the soil mixing with the sweat on his hands, he said, "A beetle pollinates the ginger, which is a good medicine." Then he moved off to show us blue cohosh, black cohosh, pawpaw, crow's foot, stinging nettle, and lady's thumb.

At the highest spot on Mr. Morgan's land, we climbed up a dirt mound. Straight below us, the bulldozers had left behind a mass of discarded trees and wasted earth. A slurry pond pooled on a shelf where the machines had torn open the mountain. The humid air cast a blanket of mist over the mountains beyond. One author said, "I don't imagine it'll be too long before those go, too. Better get a good look."

At the Morgans' home, a feast awaited us. Fried chicken, pasta salad, potatoes, green beans, sandwiches, and pies packed the dining room table. After lunch, I gazed out at the strip job in Mr. Morgan's front yard. The machines' beeping, like the beeping of a garbage truck, made me see how Mr. Morgan was right in saying he'd been shoved back in a corner. Those mountains that were a part of his history were coming down around him.

We drove up to a lookout spot, where passing coal trucks flung dust at us. I gazed out at miles of flattened mountains, a desert plateau reaching to the horizon. Nothing with the dignity of a mountain remained. Except for a few islands of coarse non-native grass produced from hydro-seeding (a water-fertilizer-seed-mulch mixture sprayed from a truck) the mountains were totally denuded, left in pathetic plateaus where the dirt turned gray in the sun and mixed with sediment.

It was almost funny when Mr. Morgan jumped off the road-side berm and hurried to pull a poke plant out of the hydro-seeded grass. He carried it like a treasure for us to see: one plant species had survived mountaintop removal. My eyes darted between the chopped-up mountains and that handful of poke, all that remained of the original glory.

We walked along the reclaimed ridge where Mr. Morgan grew up and which he affectionately calls Huckleberry Ridge, and he reminisced about his childhood home. "I was raised right over here on the fork," he said. "It's called Deadening Fork. Don't ask me what gave **The beech trees are gone. At one time, this was rich land, but now all that grows is scrub.** it that name... About all the mountain has been tore down. Even the way they build 'valley fills' is wrong. They're supposed to segregate the top soil and put it back, but they don't." The companies use explosives to rip the tops off mountains, then they dump the resulting rubble into nearby valleys. Mr. Morgan pointed. "I used to hunt back in here. We tended the corn in here, too. We didn't have any fertilizer but we didn't need any. Now that old apple orchard is all tore up. The beech trees are gone. At one time, this was rich land, but now all that grows is scrub."

The mining also buried the stream where he fished as a boy. Mr. Morgan recalled that whippoorwills, Kentucky cardinals, wrens, catbirds, bluebirds, woodpeckers, wood thrushes, grouse, and sparrows once inhabited the area. Today, they're gone.

A massive boulder sat at a bend in the road, launched in one of the explosions used to level mountains. "You'd never believe there was anything powerful enough to blow those mountains all the way over here," he said. Later, I'd learn that just such a stray boulder, dislodged during the illegal widening of a coal-haul road in August of 2004, had killed a three-year-old boy who'd been sleeping in his crib near Appalachia, Virginia. The coal company appealed the $15,000 fine for the boy's death.

As we drove to the Hazard airport, we saw a few "reclaimed" areas. Reclaimed areas are parcels of land that coal companies have supposedly returned to a higher and better use after they level the mountains. Bill Caylor, president of the Kentucky Coal Association, has said that by reclaiming mountaintops, they're creating land for "sustainable development for future generations."

The reclaimed areas I saw included a mobile home park. We also passed near the Hazard ARH Hospital, which was built on a valley fill—a former valley now filled with rubble. The hospital had closed its cafeteria because it was sinking in the rubble. We heard about a Martin County federal prison built on a valley fill. People call it "Sink Sink" because an extra $40 million was spent on-site, trying to stabilize the ground, yet the guard tower is still sinking. I saw no elk or cattle grazing on the hydro-seeded areas. I also saw no golf courses or malls, attractions that the coal companies claim will draw tourists to Eastern Kentucky's "reclaimed" mountain tops.

At the Hazard airport, Dr. Alice Jones, an Eastern Kentucky University professor, spoke to us about coal mining's impact on waterways. She explained that every day, mining directly affects 750,000 Kentuckians through the water they consume from the Kentucky River. She explained how mining changes the hydrology of a normal forested environment, causing more frequent and severe flooding, while releasing sediment into waterways. According to her, when chunky gravel from mining fills a catchment, sediment becomes the Kentucky River's number one pollutant and an expensive item to filter from drinking water. Sediment also interrupts the life cycle of organisms at the bottom of the food chain, killing mussels and aquatic life.

After the lecture, I climbed into a tiny plane. The plane's doors were thinner than a car's. I swallowed hard and buckled my seat belt. Coal company executives tout the Hazard airport as a model of good reclamation, but from the air, the airport disappeared amidst the mining wasteland.

Everywhere we looked the earth was torn apart. A massive slurry pond appeared as a gray blue mess. In the distance I saw a mountain explode in a volcano of translucent particles.

A photographer in the plane asked, "Could I get a contrast photo of a pristine spot?"

The pilot laughed. "You'll have to zoom in real close. Everything's been logged or mined." He was right. The only trees were just a few patches around people's houses. To my eyes, Perry County could no longer be considered a part of Appalachia. The mountains were gone.

After landing, we headed to the Hindman Settlement School, a cluster of wood-framed buildings on a hillside. There we would have dinner with local people, and hear the testimonies of coalfield residents at a community meeting.

John Roark, a resident of Montgomery Creek in Perry County who'd worked in the coal industry for thirty years, ate dinner with me. He was adamantly opposed to mountaintop removal. "When they start blasting real heavy, it jars the ground and the bedrock," Roark said. "We have a lot of iron and sulfur that it dislodges into the water. Plus, you get cracks in your wall and your foundations. Then when you remove the trees and the vegetation for mining, you get flooding. The low cost of energy is not low cost if you live in Eastern Kentucky."

After dinner, we moved to the adjacent room for a public hearing where the writers formed a panel to listen to the audience's testimonies. Erica Urias, the first person to address us, said that she was scared to bathe her baby daughter out of fear that the child would put the poisoned bathwater into her mouth.

Ernest Brewer of Hueysville said he takes his children outside to play knowing they'll return "looking like coaldust." He told us about the blasting that occurs every night at 11 p.m., 1 a.m., and 3 a.m. "If I'd known that the coal company would move in two months after I did, I wouldn't have put $89,000 into that house. I would've put it into a U-Haul trailer."

His neighbor Clinton Handshoe described how conditions in the community made him feel like a prisoner in his own home.

Another woman was infuriated because mining had destroyed a section of her road, preventing ambulances and school buses from getting through.

John Roark echoed her sentiments. "We should have the best roads, the best schools, with the money that goes out of here," he said. "But that money goes to someone in Miami or someone in Boston, and they never put any money back in the community."

We also heard testimony from a woman whose 21-year-old daughter had been killed when an overloaded coal truck hit her car. On that one-mile stretch of road, four victims had been claimed by coal trucks. For her, the coal trucks made U.S. 23 such a dangerous highway that "you can't even get your mail out of your mailbox" for fear of being run over.

Identifying a lack of effective laws and enforcement as a primary problem, Letcher County Judge-Executive Carroll Smith said, "There's the case of a house that had one foot of mud in it, yet the inspector defended the coal company. Why? Because he couldn't find anything to write the coal company a citation for. Neither the state nor the federal government will find fault with a coal company for sinking a house. Any law that allows a company to flood a person's home is *wrong*."

The last speaker of the evening summed up the dilemma that so many Eastern Kentuckians find themselves in: wanting to protect the very mountains that their neighbors are employed to mine.

"People are afraid to say there's a right way and a wrong way to mine coal," she said, her voice shaking with emotion, "because fifty of your neighbors will kill you for threatening the way they make their money. By the time we do something here, these mountains will be gone."

"God made these mountains," she went on, "and nobody can ever put them back to their original nature. You take a look at that stuff and you tell me how one thing will ever grow on it. It's a lie."

"It's easy to take our coal," she went on, her voice filled with tears of anger.

"We're drug addicts, we're idiots, we're poor. If the people in California cared about this as much as they care about the Arctic Wilderness . . ." her voice trailed off.

"The world owes a debt to Eastern Kentucky that they can never, never repay."

Blasted Away

by Charles Bracelen Flood

This spring my daughter Lucy and I received an invitation from Kentuckians For The Commonwealth to join other writers in a trip to see the effects of mountaintop removal. Off we went: I, aged seventy-five, the author of eleven books and the oldest person to be in that group, and Lucy, a 25-year-old writer of short stories and essays, the youngest. During the next two days, we occasionally talked together, but spent most of our time just looking and listening.

My first powerful reaction came after walking through a beautiful wooded area in Perry County. As we came to the top of a massive slope covered by a marvelous variety of plants and trees, I was filled with the serenity than comes from spending an hour in that kind of harmony with lovely surroundings.

That changed in an instant. Coming to the crest of that wooded ridge, we looked straight down into what should have been the other half of that hill, but it had been scooped out into an enormous slanting expanse of broken rock and yellowish and gray gravel. Hundreds of feet below us, bulldozers charged back and forth amidst thick clouds of dust, tearing more coal out of the guts of this mountain. This was "mountaintop removal"—a remarkably sanitized term for what I was seeing. From time to time, there was a blast as dynamite tore apart large parts of the earth, the "thump" coming to us from below at about the same moment that our feet felt the impact.

That afternoon, our group had the opportunity to look at a far more panoramic view. We were driven to the Hazard airport, and, in threes, were taken up in light four-seater airplanes. I have traveled on five continents and been a war correspondent, and I have never seen anything like what I now beheld. The landscape below us was a patchwork of beautiful hills in their spring greenery, and other hills torn to shreds. The devastation was grotesque. The bare slopes bore no resemblance to the erosion one sees in the Grand Canyon, in which there is some logic and beauty in what a river has cut out over countless millennia. No beauty here: if one half of a hill has coal and the other has none, the bulldozers drive off, leaving scooped-out slopes standing at ugly angles that can only be produced by human actions. Below me lay a wasteland. Rising from this manmade desert were grey clouds of dust; I coughed every few minutes for two days after that plane ride.

The mutilated condition of the mountains spoke for itself, but that night we heard from the people who live in the area. At the Hindman Settlement School, we gathered after supper to hear from a succession of local residents. One wrenching story followed another. I was struck by the account of a man whose small plot of land was being approached by one set of bulldozers coming up from a valley, and another set working their way down from a ridge above him. Technically, no one was trespassing on his property, but his house was constantly being shaken by blasting, and mudslides caused by that blasting were coming down around his property.

It was somewhere during this evening that my shock turned into anger. These were American citizens, effectively being deprived of all potential for a decent life. They are victims, their rights blasted away as surely as the contours of the hills in which they live. The politicians whom the coal companies support are silent when it comes to enforcing regulations that would stop the wrecking of the world in which Eastern Kentuckians are forced to live.

That night we slept at the Hindman Settlement School. I got to bed earlier than the younger generation, and lay there listening to the voices talking in a big nearby living room, the voices of gifted young writers who will be fighting for justice for those who are being exploited, and for the preservation of a beautiful countryside that should not be slashed to ribbons.

The next morning, after breakfast, we gathered to share our thoughts. A concern of mine, then and since, is how to make Kentuckians who live far from those hills understand that we are all in this together. All this blasting in the hills is polluting our streams at their very sources; the quality of water in the Kentucky River is deteriorating. No fewer than 750,000 Kentuckians depend on that river wholly or in part for their water supply. I would like to say this to the lady in Danville, the legislator in Frankfort: It's coming your way, folks. You either reverse this trend now, or pay for it in your health and in great numbers of tax dollars in the very near future.

As for how the folks that are causing these threats feel about themselves and those who dare to speak against "mountaintop removal," right at the time of our tour there came a statement from Mr. Bill Caylor, speaking on behalf of the Kentucky Coal Association. He said, among other things, that "mountaintop-removal mining affects a very small percentage of the mountains" and that it "leaves valuable flat land for future generations." We saw some of that "valuable flat land" on our way out of the mountains. A mall had been built on a mass of the crumbled rock and gravel that results from the dynamiting, and part of a large retail building at one end of the mall had been closed because it was sinking. As for that "very small percentage," there are counties in West Virginia in which 40% of the hills have been irretrievably scarred and made unlivable for human beings or animals.

Mr. Caylor also had something to say about the group of writers who had invaded the coal companies' fiefdom. He said of us, "These are the same people who would be outraged if they knew where their ground beef came from." As part-owner of a farm on which we raise beef cattle, I know the difference between traditional agricultural practices and the one-time-and-move-right-on permanent destruction of some of the most beautiful scenery in the United States.

I think that it may indeed be up to people like those young writers to spread the word about what they've seen and heard. It's a funny thing about writers, Mister Caylor. When they get into something like this, they don't know how to quit.

They've knocked me down, all right—
but I always got back up on my feet.
And I swear I won't fall again.
These mountains can't
protect themselves.
We got to do it.
I got to do it.
—Wayne Bernard

18

Fossil Fuel

by Davis McCombs

We found cars and semis burrowing
the air up the hill's steep grade,
wind off their grilles and mudflaps
dragging snow in bands
over the pitted asphalt, but not
the Trilobites we'd come for.
I thought of sleeping in a bed
of shale, undisturbed by the chisel's
sulfurous clank or the seams
of moisture that, freezing, wedge
the bluff's cracked joints apart.
A hammer, a road cut, a tourniquet
of vines: We didn't find a thing
worth taking. The day picked
the flesh off the hill's white bone,
tightened its talons of sleet
on the rock. I hunched my back
to the wind and wintry mix, shivered
in my thick coat, and set my face
toward a temple built in air:
that great downshifting wake
of traffic that would leave,
I knew, not one stone upon a stone.

Statement on Mountaintop Removal

by Artie Ann Bates, George Brosi, Dexter
Collett, Charles Bracelen Flood, Lucy Flood,
Silas House, Kristin Johannsen, Loyal Jones,
Ed McClanahan, Ann Olson, Erik Reece,
Gwyn Hyman Rubio, Anne Shelby, Bob Sloan,
Mary Ann Taylor-Hall

The following is the statement issued by a group of Kentucky authors after their tour of mountaintop-removal strip-mining sites in Leslie and Perry counties, in conjunction with the Kentuckians For The Commonwealth:

Yesterday we witnessed appalling destruction to the land. The practice of mountaintop removal to extract coal is ravaging Eastern Kentucky, and its effects are headed your way. Mountaintop removal represents economic and cultural violence which eventually reaches the whole state. What we have seen has convinced us that mountaintop removal is a blight on the entire state that is robbing our people of a better future by destroying our most abundant resources and the very ones we will need for building a viable future economy. Streams and groundwater, scenic beauty, diverse forests, and native plants are all being ruined forever by mountaintop removal.

During our two-day tour of mountaintop-removal sites in Eastern Kentucky, we saw buried and polluted streams, great hickories and oaks tossed into useless piles, life-giving mountains turned into barren moonscapes, wasted topsoil and sunken homes, the lowering of a people's quality of life, the increased severity and frequency of flooding, the lost jobs and lost hopes of an entire place. These are not isolated or occasional incidents. Instead, they are an assault on the people, culture and land of Appalachia.

We have met these people and heard their testimony. We learned of a one-mile section of road where four mothers grieve for their dead children, victims of speeding, overloaded coal trucks. Erica Urias spoke eloquently of bathing her baby in poisoned water. Clinton Handshoe, surrounded by strip mining's noise and air pollution, referred to himself as "a prisoner in his own house."

We realize that coal is an important part of our economy. However, coal can be mined in a more responsible way that respects the spirit of the land and its people. Out of greed, we have forsaken moral, aesthetic, and spiritual values. We have traded the futures of our children and grandchildren for cheap coal. The impact of these practices is sweeping across the entire state faster and faster, through the spread of air and especially water pollution.

We are horrified that this practice is legal. We are angry that representatives in our own government are allowing this to happen. Mountaintop removal is not right, it is not acceptable, and it is an act we will fight. We call for the abolition of mountaintop removal and urge our fellow citizens to pressure elected officials in every way to stop this criminal desecration of our common wealth.

See for Yourself —

Viewing the photographs and reading the descriptions of mountaintop-removal devastation in this book cannot adequately convey to the reader an understanding of the extent of destruction of Kentucky's mountains. And many will never have the opportunity to visit these nearly inaccessible areas, or view them from the air. However, thanks to the internet, one can view satellite photographs of the region described here. Google Maps www.maps.google.com has a feature which overlays maps of highways over satellite photographs. These photographs from space clearly show the devastation. For example, simply search for "Ary, KY," then select "Hybrid map," next, zoom out a bit and observe the region east of Hazard along Highway 80 between Hazard and Hindman. There and especially farther on to the northeast you'll see mountaintop-removal sites too numerous to count, adding up to many thousands of acres.

Stripped

by George Ella Lyon

I was humming "Mist on the Mountain"
and shelling peas
I was figuring board feet
I was carting off stones
and quilting lettuce
and thinking about a baby growing ripe inside
I was voting
I was lifting pain out by the roots
the bread indoors
breathing beneath a thin towel,
when the D-10 dozer came
and rolled me off the front porch.

Bringing Down a Mountain

by Stephen George

Two convergent myths figure into how the rest of the world perceives Eastern Kentucky. One says Appalachians are poor, illiterate hillbillies with drug addictions, fated to poverty. The other, more romantic storyline, tells of people with a profound connection to the land and a spirit guided by the trees and mountains, a narrative introduced to the larger world through bluegrass music and the writing of Wendell Berry, Gwyn Hyman Rubio, Silas House, and many others.

One facet runs through both myths like a central theme: Coal. It powers nearly everything Americans take for granted—lamps, refrigerators, even iPods. Most things that run on electricity come to life because of coal, and yet it's hard to imagine that most of us ever give that a thought. And because coal powers more than half of the United States and much of the rest of the world, the coal-rich mountains of Eastern Kentucky are a major supplier for the coal industry.

At its peak, some 50,000 Eastern Kentuckians worked in the industry, but that figure is now more like 15,000. Likewise, the symbols so connected to coal over the years have changed. The stark contrast of a miner's white eyes against his dust-covered face has given way to a front loader; boots and overalls have been swapped out for the image of leveled mountains.

Before the mid-1970s, coal miners went into the ground to harvest coal. But since the stealthy beginnings of mountaintop-removal mining in the 1970s, miners have become nearly irrelevant, replaced by massive amounts of explosives and heavy equipment. In mountaintop-removal mining, workers literally blast the tops off of mountains with explosives, then harvest the coal. The practice has turned nearly seven percent of Eastern Kentucky's lush landscape into something more closely resembling the bare and dull-gray terra of the moon. In a matter of weeks, tree-filled mountains turn to plateaus. Hollers where headwater streams once flowed are filled with the old mountaintops, rock and soil that coal companies call "overburden." These valley fills can bury water supplies for entire ecosystems.

And the results are ruinous. Most who live in and around the mountains get their water from wells, which are at risk for contamination when acidic drainage runs off into the head-water streams. Flash flooding—a fact of mountain life forever, considering the topography—occurs on a larger-than-natural scale, which is to be expected when you remove so much of the landscape that's vital to controlling rain runoff.

Coal companies refer to substantial unnatural flooding as an Act of God.

Coal companies refer to substantial unnatural flooding as an Act of God, a semi-legal term that exempts them from financial or other responsibility when, say, a family's home takes on five feet of water and half of its possessions are destroyed.

Jonathan Phillips is a professor of geography at the University of Kentucky with expertise in hydrological issues related to mining. He recently completed a study in Eastern Kentucky examining flooding problems around mining operations.

"Since the only rainfall records kept up in those remote areas are by the coal companies themselves, and of course they're not out monitoring places where they're not mining, it's very difficult to test," he said.

Phillips also said it's unlikely, despite the typical coal company line, that a rainstorm would bring "significantly more rainfall in a filled valley as opposed to a nearby unfilled valley." Yet flooding around the mining operations has become a serious problem.

Pleasing the King

King Coal—a moniker that refers to the amalgamation of coal companies, the Kentucky Coal Association (KCA), and most of our state's coal-friendly politicians—will not acknowledge that mountaintop removal is destructive, either to the environment or to those who live near mine sites. Bill Caylor, president of KCA, does concede that mountaintop-removal sites look bad while mining is in progress.

"It looks ugly," he said. "It looks like a lunar landscape."

But that's about as far as he'll go. Caylor describes mountaintop removal as "sustainable development," a reference to reclaimed mine sites where things like golf courses and pastures are developed where mountains used to be.

"In fifty years, you will have use for this land," he said. "You can't use the steep hillsides."

Such a view typifies the bizarro, Alice-in-Wonderland through-the-looking-glass logic that comes into play in such discussions: asserting that Appalachian tourism cannot revolve around the natural beauty of a prehistoric mountain range, but that it will thrive around things like golf courses on leveled plains, is the opposite of logic.

The fact is, a lot of money follows King Coal around, and most of it walks right out of the state with him. Kentucky ranks third in national production of coal, behind Wyoming and West Virginia, but 80% of the region's coal leaves the state, according to KCA. It is a multibillion-dollar international industry—KCA reported that coal brought $2.5 billion into the state in 2000. But suggesting that money is fed back into the economies of the counties where coal is mined reveals more dyspeptic logic. Two companies that command a major presence in the region are based outside the state. Massey Energy Co., whose subsidiary Martin County Coal was responsible for a 300-million gallon slurry spill in 2000—thirty times larger than the notorious Exxon-Valdez tanker disaster—is based in Richmond, Virginia. TECO, another considerable presence, is based in Tampa, Florida.

And Appalachia remains among the poorest areas of the country, while Kentucky has the lowest energy prices in the nation. Kentuckians spend, on average, one-third as much as someone living in Hawaii, for example, to power his or her home. Coal provides all but three percent of Kentucky's energy.

It is hard to imagine that the average person would not trade just a little convenience—a light off here and there, TV on only in the room where it's being watched—to reduce the environmental toll that coal takes on Eastern Kentucky.

Here Comes the Boom

Mountaintop-removal mining is a simple process: plow the trees (but don't bother harvesting them) and everything else living on the mountain, blast off the top (usually 800 to 1,000 feet), take out the coal, and leave a leveled area that, under the Surface Mining Control and Reclamation Act (SMCRA) of 1977, should be restored as closely as possible to its original contour or to some otherwise usable form.

It may be helpful to imagine a mountain as a layer cake: at the summit are trees, grass, thickets and other undergrowth. That would be the icing. Under it is earth, the first layer of spongy cake. Next, there's a thick streak of ice cream. That's a seam of coal—or more than one. There are lots of them down the center of many of these mountains. In fact, according to KCA, some 84 percent of the region's original coal deposits remain. Eastern Kentucky won't soon run out.

After a permitting process that may take two years, workers plunge a drill as wide as a basketball sixty feet into the soil. The holes, every ten feet or so around the mountaintop, are then filled with AMFO—ammonium nitrate and fuel oil. That's the same chemical cocktail Timothy McVeigh used to blow up the Oklahoma City Federal Building in 1995. According to an essay by

Erik Reece for Harper's magazine, McVeigh needed 4,000 pounds to remove the side of that building; the typical mountaintop operation may use in excess of 40,000 pounds. The mixture is so volatile that the two chemicals are delivered to sites separately.

Remotely wired caps are placed on the holes. Sirens sound three times to warn the neighbors. Then … BOOM! Thousands of tons of rock, earth and sediment fly indiscriminately as the top of the mountain crumbles. Trucks dump the overburden into valleys, the hollers between peaks that house not only families but headwater streams as well.

The stark disparity between the industrial complex atop these hills—haul roads, growling coal trucks, all the dust—and the natural roll of the remaining green mountains is striking.

Grass growing between the rocks on a reclaimed strip-mine site.

I Fought the Law, but …

According to the Surface Mining Control and Reclamation Act (SMCRA), the 28-year-old law governing such activities, companies that mine a mountain must return the land as close to its original contour as possible. A stipulation in the law, however, allows a reclaimed mountaintop-removal mine site to be leveled if it is put to another use. So coal companies are creative in how they rework former mountains; some areas they call pastures, and a few wild horses have been left atop one site, according to the *Lexington Herald-Leader*. Another designation is "recreational area." In Perry County, there's a spot with a bench atop a swath of bald, gravelly land. It affords a delightful view of a several bald plateaus.

More commonly the sites become residential areas. Along the road leading up to the Wendell Ford Airport in Perry County—itself built on a reclaimed mine— mobile homes are scattered, with dirt for yards and front doors within twenty feet of trucks carrying 100,000 pounds of coal, barreling down the mountain.

When he took time to help the man up the mountain, he scaled it himself.
—Anon.

The point is, what once was the exception to SMCRA now seems like the rule. Mountaintop-removal mining is perfectly legal. So are valley fills and half-baked reclamation sites. Judging by what's been written, it's difficult to persuade folks that there's anything distasteful about that.

Tom Fitzgerald, the Louisville attorney who is prominent in statewide environmental issues, is working with Kentuckians For The Commonwealth (KFTC) to change the rules governing where coal companies can dump overburden. He authored House Bill 509 for the General Assembly's last legislative session. It would require coal companies to return overburden either to old sites that have not yet been restored to their original contour, or to areas determined during the permitting process. It would also eliminate dumping in any Kentucky waterway. The bill failed to get a committee hearing, but Fitzgerald expects it to advance in the next session.

The strategy, in effect, is to force companies to move their debris from mountaintop-removal sites to other locations, a practice which would increase costs. Whether that would be enough to force companies to reconsider their actions is debatable.

Five Democrats co-sponsored HB 509: Don Pasley, who represents Clark and Madison counties; Reginald Meeks and Jim Wayne, both of Jefferson County; Harry Moberly, Madison County; and Kathy Stein, Fayette County. According to Fitzgerald, the five legislators are the only members of the Kentucky General Assembly who've supported legislation that would further regulate coal waste disposal.

"Yeah, it is, it's difficult," Fitzgerald said of passing legislation that affects coal companies. "Particularly when it's legislation that would impose additional costs on the coal industry relative to the handling of their waste material."

Part of that difficulty arises because the industry feeds a lot of money into the political process. In 2004, for example, the Center for Responsive Politics reports that coal companies donated more than $2.3 million to candidates nationwide. Ninety percent of that went to Republicans.

It's a seemingly impenetrable union of politics, policy, and corporations. For instance, the Office of Surface Mining, which handles complaints about mountaintop-removal sites from citizens, answers to the Department of Labor. That cabinet is headed by Elaine Chao, who is married to Kentucky's senior U.S. Senator, Mitch McConnell, who has accepted millions from the coal industry during more than two decades in Congress. McConnell is known for the axiom that "money equals free speech."

Rep. Wayne, who has previously worked with KFTC on tax reform issues, frames mountaintop removal in moral terms. While the coal industry references all decisions to cost, people who live in mining areas pay a price with their health.

"It's destroying lives, it's destroying communities," Wayne said. "It can't continue the way it's being done. If that means imposing restrictions on the coal companies so that they can no longer do business this way, I'm OK with that."

There is, Fitzgerald says, at least one way people who are affected by mining can counter the SMCRA loopholes. Companies must apply for separate permits for each mountain they want to mine. Different operations require different permits, however, so many coal companies apply for, and are granted, nationwide permits that basically allow them to conduct any operation, regardless of scale, at the location for which they've applied. Fitzgerald and others refer to these as "drive-by permits," because they are simple to obtain.

For large operations like mountaintop removal, any citizen may ask federal agencies to require companies to apply for an individual permit, which carries a much more stringent set of qualifiers, sometimes including an environmental impact statement and a "fairly rigorous review," Fitzgerald says. The process also requires companies to propose alternatives to locations where they plan to dump debris, something lacking in the nationwide permit process.

In January, KFTC and two other environmental groups filed a lawsuit against the Army Corps of Engineers, alleging that the Corps is illegally allowing coal companies to gain access to nationwide permits, which don't require public comment. The suit covers about forty new mining operations, says Kevin Pentz, a KFTC staffer. No judgment has been rendered, but that may happen within six months, he says.

"The tools are there," Fitzgerald said. "The problem is, the tools must be utilized."

Dimming the Lights

Exploitation is another built-in part of the folklore of Appalachia. Forty years ago, President Lyndon Johnson stood on a porch in Martin County and declared his War on Poverty; his words still ring hollow through the hollers of Eastern Kentucky. It appears politicians have abandoned this area while the coal companies have left leveled mountains behind, and no one seems to think a thing of it.

Considering how reliant this country is on coal to make that socket in your wall work properly, it's unrealistic to call for an end to mining coal. That's not what anyone wants in Eastern Kentucky. They're used to mining; most have raised families on King Coal's dime, and everyone I spoke with was well aware of the necessity to mine these mountains.

All they're asking is that King Coal assume some responsibility in mining practices, a request that implicitly eliminates mountaintop-removal mining. They want respect—for their land, homes, and way of life.

In the Louisville area, we complain about Louisville Gas & Electric hiking prices another 15 percent, outraged that our heating bills are peaking well over $100 for modest-sized apartments. The disconnect between our city lights flicking on at the touch of a switch and a headwater stream being buried by a valley fill in Perry County is profound. Yet the two are inextricably linked: energy comes from somewhere, not just from the wires in the walls. And it will continue to come from coal. President Bush's recently unveiled energy policy includes proliferation of coal-fired power plants for increased energy production here at home.

Upward growth and major reliance on such resources as coal parallels the declining health of the planet. According to the National Academy of Sciences, the overwhelming increase in carbon dioxide deposits into the atmosphere—a result of burning coal—has left us with an average global temperature that by century's end will be hotter than at any time in the last two million years. The result: the United Nations' Inter-governmental Panel on Climate Change reports that over the past 100 years, Earth's sea levels have risen about six inches, because of melting glaciers and ice caps.

But with fatalists and fundamentalists filling Washington's glitziest (and most influential) government offices, fervently believing the Rapture is nigh, a focus on environment is pretty much a moot point and the argument for self-preservation falls upon deaf ears.

Appalachians are used to deaf ears. They have learned to live a different way, so that they may sustain themselves and their land long after the building lights go dim. But now, their natural legacy—a splendorous spread of rolling hills and green mountains mirrored nowhere in the world—is being systematically destroyed so that an unsustainable way of life in our cities may continue.

Truth Buried Under Natural Material

by Anne Shelby

Maybe we aren't being fair to the coal companies, I thought, after I got home from the Kentucky authors' tour of mountaintop-removal sites and had a few days to think things over. Maybe the coal industry isn't the Big Bad Wolf here, and maybe those of us on the tour were, as Kentucky Coal Association President Bill Caylor said in the newspaper, a group of educated, well-meaning, biased people who don't know where ground beef comes from.

I decided to check out the coal industry's side of the story, and began by visiting the Kentucky Coal Association (KCA) Web site.

There, I learned that the coal industry is the victim of "false and/or misleading impressions," and that the Kentucky Coal Education Project (KCEP), a prominent link on the KCA Home Page, aims to correct this injustice by offering "viewpoints seldom completely represented in the news media." (All quotes are from the KCA and KCEP Web sites.)

The KCA site was indeed educational. I learned that while companies sometimes "mine" or "extract" coal, more often they "recover" it, and that mountaintop removal is the most efficient way to "recover" this coal, which the industry apparently misplaced in the earth several geologic ages ago and is only just now getting back. I learned that, contrary to popular belief, the Martin County slurry spill was not toxic, but consisted of "simply water, coal, and rock particles" which were "all natural material."

In the comfort of my own home, I could choose from a list of informative articles with titles like "Top 10 Global Warming Myths," "The False Promise of Hydrogen," "The American Lung Association's Phony Air Scare," "Coal's Romantic Side Keeps Peabody Chief Passionate," "What Kills People During Air Pollution Episodes?" and a long essay on mountaintop removal by Kentucky Coal Association President Bill Caylor.

Caylor begins his article with modest, even ambivalent, goals: "We feel that it is important to accurately present another perspective on mountaintop mining so that students can examine for themselves the various viewpoints and arrive at their own conclusions on this mining method. What most individuals and students will learn is that there is never any clear-cut solution or answers to many of the complex problems facing today's society."

But by the end of the article, any ambivalence or acknowledgment of the complexity of the issue has vanished like morning fog, and Caylor has convinced himself that mountaintop removal is "simply the right thing to do—both for the environment and for the local economy. A true win–win."

Caylor reaches this surprising conclusion by responding to some of the most often-heard criticisms of mountaintop removal, or, as the industry calls it, mountaintop mining. (The word "removal" has been removed and buried under natural material.)

The article offers a timeline, which begins with August 3, 1977: "Mountaintop mining existed in Kentucky. Where have all these activists been? This mining method isn't new. The use of valley fills isn't new. The only thing new is this attack on a method of mining that's been practiced for well over 20 years." And so we are off on our informative, dispassionate overview of the history of mountaintop mining.

To the charge that mountaintop mining is destroying Kentucky's streams, the Coal Association president explains that streams are not actually destroyed, lost, or buried, but merely "rechanneled." And besides that, the argument continues, they aren't even really streams. They "basically flow only in connection with a rainfall event. We would rather refer to these streams as 'gullies.'"

Mountaintop mining, it turns out, is actually good for the environment. And not only that. According to the KCA, mountaintop mining is "one of the keys to the economic future of Eastern Kentucky. . . . This mining practice creates land that has the potential ["potential" is underlined on the Web page] for many other uses." Examples: farming, growing trees, creating sites for housing developments, individual homes, airports, golf courses, schools, industrial parks, recreational areas, hunting lodges, shopping centers, correctional facilities, the much-needed development of alternative agriculture, soccer, baseball or football fields. Mountaintop mining also provides ponds, wide roads, and "spectacular views."

Caylor admits that not everyone who looks at a mountaintop-removal mining site can see these things. "You must have long-range vision," he concedes, "to see the dramatic positive changes that will occur to those areas with level land."

With this special vision into the future of Eastern Kentucky, Caylor does not see tourism as part of the picture. This is not, as one might think, because tourists are unlikely to want to view decapitated mountains and deforested hillsides. In fact, this concept does not appear in the article at all. It is, rather, because tourism, though "touted by many as a viable alternative . . . is principally built on minimum wage jobs and . . . is no bargain for a region . . . Ideally, factory wages are the goal."

What Caylor sees in Eastern Kentucky's future is not tourism but factories, which will be built on the level land left by mountaintop mining. Presumably American industry, having deserted its manufacturing centers in the North, is now traversing the globe in search of flat land, and will return to this country when enough mountaintops have been removed and enough valleys filled.

The article does not answer the criticism that mountaintop mining buries or erodes topsoil, that thin soft black layer upon which life depends. This was perhaps an oversight. In the spirit of cooperation and mutual understanding fostered by the Coal Association Education Project, I would like to suggest the following addition:

> In mountaintop mining, the topsoil, or what we in the industry refer to as "non-coal material," is not lost or buried or washed away, as some overly emotional environmental extremists like to claim, because nothing in Nature is ever really lost or destroyed. The topsoil is simply relocated to areas under the mountain, where it is protected by many layers of rock and other natural materials, or it is transferred further downstream by means of rainfall events to rivers, lakes, oceans and other pleasant places.

The Coal Education Project web site is funded by something called the Kentucky Foundation, and the site also acknowledges "state funding of this project." Does this mean our tax dollars have been helping to pay for coal industry propaganda—I mean, education information?

Here's part of what the KCA Web site says about Association President Bill Caylor: "He knows the political process—how the General Assembly and the various regulatory agencies work—and how to impact it." And that particular piece of information, I feel quite sure, is incontrovertible fact.

Red Raspberries

by Artie Ann Bates

On summer evenings, when dense vegetation covers the scarred terrain, the mountaintops almost look present again. Our mined-out place is quiet—I no longer hear the beeping back-up sound of trucks and loaders, or feel the blasting rattle. My dogs come with me on our walk to the red raspberry patch by the road. If we hear a car, we step back so that no one can see us and discover our mission. As I eat berries and they chase violet butterflies, I sense nature moving slowly, healing at her own pace. Sometimes she plants us, like the berries, in places where we can help.

At the end of my gravel road leading to the one-lane paved road that runs three miles out of Elk Creek, the red raspberry bushes sprang up this past summer. The seeds were no doubt carried by birds from a small bush up a nearby gas road. The berries got their fullest ripe by mid-July, the hottest time of the summer. At first, practically no one knew about them, so I had them all to myself, as if God had treated me alone. Now others come to pick, but I still go there and pretend that no one else knows. The berries are deep reddish purple, bursting with sweet syrup that sticks to my fingers. I reach further into the bushes without caring that tiny thorns pierce my arms. My fear of copperheads in the low weeds pales as I reach the highest, fullest berries, ones that usually only the birds can capture.

My mother's grandparents lived on this holler farm in Elk Creek from 1907 until their deaths in the 1920's and 30's. My dad was raised just up the holler. Daddy recalled the 1927 flood when he was eleven years old, with lightning showing whole trees, still upright, washing down the holler. In 1956, my family moved to this farm; at age three I ran across the yard in the evenings to meet dad as he came home from work with ice-cream sandwiches. That was before the changes.

One surprising day in the late 1960's, I came home from high school to see two men in suits sitting in our living room with a broad form deed spread out on the table. Daddy was no match for those rogues with dollar signs in their eyes. They tasted blood. Being a product of Eastern Kentucky culture, Daddy didn't know how to be rude to strangers. But years later, in a huge victory, he kept them from dumping overburden onto our historic farm. By the mid-1990's, all the mountains surrounding our log homeplace had been lowered by at least 150 feet. The companies would mine one vein of coal, then come back a year or so later and mine another. This went on for fifteen years.

When the world falls down around us, and we see the Creation destroyed, we cling to small things of beauty for healing, like red raspberry bushes. These mountains, once as tall as the Rockies, have eroded naturally for over 200 million years. Humans have lived around Elk Creek for more than 9,000 years. They left few traces, until the advent of European settlers 200 years ago and the process of industrialization that came along behind them. The changes happened faster and faster, most violently and irreparably in the past thirty years of mountaintop mining. In my life of 52 years, these mountains have lost more height and body than in the previous several million years. My son's generation will never know them as they were, nor witness the flow of human life up this holler.

What my son's generation does know is that trying to live in Eastern Kentucky means having extremely limited job opportunities. In this state, which is third in the nation in coal production, with 75 percent of it coming from the mountains of Eastern Kentucky, there is little work except in education, health care, and coal mining. Since roughly half of the mining jobs now available are in mountaintop-removal operations, young people are forced to choose work that requires them to destroy their own land. Much like a recently stranded rock climber who had to consume his own hand as a source of food, surface miners have to wreck their children's future to live here today and put food on the table. What a choice.

. . . the natural world sustains humankind if we respect it. As long as we do not destroy nature, she will replenish our food and water supply.

And the choice has made us contrary. We are intense yet soft-spoken people, easily angered underneath our sense of humor. We are generous to a fault, but suspicious of foreigners. We have mineral wealth but many of us are on food stamps. As a people, we resist zoning but let the coal and gas corporations intimidate us. We view ourselves as independent, though many here are dependent on narcotics. While our gardens hang with the weight of beans, corn, and heavy tomatoes, we prefer Wendy's and McDonald's.

Conflicts about coal, like the conflicts about the Civil War generations ago, often arise within families. Some have brothers who work for the coal companies, and sisters who publicly condemn them. Ultimately, for many of us, the attachment to family and land trumps a better living in the city. So we stay, and try to make the best of what is left.

The Native Americans who hunted and trapped in Kentucky's Appalachian Mountains knew that the natural world sustains humankind if we respect it. As long as we do not destroy nature, she will replenish our food and water supply. But we modern Americans seem to have forgotten that. The destruction of the Appalachians by mountaintop removal is the ultimate lack of respect for the Creation. The responsibility lies with the greed of the coal companies, fed by our American addiction to electricity and cars. We be pigs. Healing the scars of mountaintop removal—including the lopsided economy, and the anger within families and communities—will take more than a patch of red raspberries. But the berries represent the richness of the natural world. In winter, the mountaintops are obviously missing. No amount of pretending will replace them. The silhouettes of trees are scraggly and weak. But in summer, the raspberry bushes strut their stuff. The summer green enwraps our wrecked hills, bringing hope to this place and its people, who have endured great loss.

The mountains, rivers, earth, grasses, trees, and forests are always emanating a subtle, precious light, day and night, always emanating a subtle, precious sound, demonstrating and expounding to all people the unsurpassed ultimate truth.
—Yuan-Sou

Forced from Home

reprinted from the KFTC newspaper

What once was home is now a nightmare for Gwen Patrick and family.

Gwen's father, Leevone Baker, shares the nightmare. He thought he had done well to provide a place for his daughter and her family to live.

That started changing two years ago. Consol Coal Company used the broad form deed to open up a deep mine on land owned by Baker in Knott County, near Beaver.

What confused Baker was that he never saw the permit application notice published in the local paper, as required by law. There was a notice, but Consol left his name off and said his property belonged to CSX, the railroad company.

Baker actually lives a few miles away. But Patrick and her family have lived just a fourth of a mile down from this new mine for the last two years.

"Two years ago they started bulldozing the hillside at 4 a.m.," said Patrick. It's been a non-stop headache ever since.

For the Patricks, it seemed as though every week was a new aggravation caused by Consol's mine. It started with bulldozing and burning of the brush from the hillside. The smoke would fill up the hollow and make it unhealthy for people to go outside.

Then slowly the creek behind their house, Isaac Fork of Beaver Creek, which is one of the headwaters of the Big Sandy, started turning blacker every day. Now, on good days, it's a dark gray color.

"My little boy use to catch crawdads and minnows in the creek behind my house," Patrick explained. "Now I won't let him get in it any more because it runs black all the time."

Patrick called state inspectors with the Kentucky Department of Surface Mining to ask that the problem be corrected. She also talked with several Consol superintendents, and they promised they would have people come out and check on the water.

One day she noticed there were some contractors taking water samples from the creek. But she never heard any more about it. They stopped returning her calls.

"It was like it was dead in the water," said Patrick.

And it wasn't just the water outside of their home that was being impacted by the mining operation. The water in their well was also turning black at times.

A year and a half ago, Consol drilled five wells for water they use during the course of mining. Baker, Patrick's father, believes Consol's wells are drawing water from the same aquifer as Patrick's well.

Consol pumps hundreds of gallons of water into holding tanks to be used when needed. But once in a while the company lets the water from these tanks run back into the aquifer. This pressurizes the aquifer and forces water to shoot up out of Patrick's well.

This water is often black with coal and silt.

Patrick says the water that runs through her washing machine is black nine times out of 10. Her family no longer feels safe drinking the water and are forced to buy drinking water.

I have been to the mountaintop . . .
I just want to do God's will . . .
He's allowed me to go up to the
mountain. And I've looked over.
And I've seen the promised land.
—Martin Luther King

State inspectors came to check whether the mine was in fact polluting Patrick's water well. The inspectors said they would put dye in the water at Consol's mine and see if it showed up in the Patrick's well.

But after the inspectors went to Consol's mine, they sent a letter to Baker saying the dye test wasn't necessary.

The constant noise and the dust from the coal mine is also a constant aggravation, enough to drive one crazy, Patrick said. The noise is so bad "you can't hear the birds in the spring . . . It's so loud, you can't go outside."

According to Patrick and her son Jeremy, the black dust from the coal settles on everything—their car, the grass, the porch. They had to wrap tarps around a storage barn because the dust was getting in on their tools.

Patrick said if Jeremy walked through the grass in their yard with white shoes, his shoes would turn black from coal dust.

One day, the Patrick's returned to their home and the dust was so thick and black as they drove up the holler, they believed their home was on fire.

The last straw for Patrick was the problem of out-of-control coal trucks on Route 7 behind their house. The trucks are loud as they go up the hill. When they come down the hill overloaded, they're often speeding out of control.

Two coal trucks in the last year have run into the ditch at the bottom of the hill, the same place Jeremy stands to wait for the school bus every morning.

"I want to move my children away from Route 7, because basically Route 7 is nothing but a coal road," Patrick said.

Not long after the mine opened, a former schoolmate of Patrick's, who now works in the mine above her house, told her, "They'll run you out. I couldn't live that close to a mine. They'll aggravate you to death."

At first she didn't believe it. After all, her father had worked in the mine. Her husband works in the mines. She's always had family members working in the mines.

But, after two years of harassment from Consol's mine, Patrick and her family feel they have no choice but to move.

"I feel like they have taken over my life. They have forced me out of my home. I have given up and we can't afford to fight them any longer" said Patrick. "They shouldn't be allowed to locate [a coal mine] this close to a residence."

"I like living here, but with this polluting of the air and our environment around the house, I think it's best for us to move."

I keep a mountain anchored off eastward a little way, which I ascend in my dreams both awake and asleep. Its broad base spreads over a village or two, which does not know it; neither does it know them, nor do I when I ascend it. I can see its general outline as plainly now in my mind as that of Wachusett. I do not invent in the least, but state exactly what I see. I find that I go up it when I am light-footed and earnest. It ever smokes like an altar with its sacrifice. I am not aware that a single villager frequents it or knows of it. I keep this mountain to ride instead of a horse.

—Henry David Thoreau

The Coal Industry's False Fronts

by Bob Sloan

Some people try to defend mountaintop-removal coal mining, the shameful practice of reducing forested hills to dead heaps of rock in order to reach thin seams of coal. It's the cheapest way to mine coal these days, partly because it requires very few workers.

These defenders of mountaintop removal nearly always mention Stone-Crest Golf Course in Floyd County as an example of how a mining site can be rehabilitated.

I never thought anything related to mountaintop removal would remind me of college history classes from forty years ago, but whenever I hear about that golf course, I can almost hear a professor lecturing to a bunch of half-attentive students about "Potemkin villages."

He told us that when Catherine was Empress of Russia in the early 1700s, she heard rumors that things weren't quite right in some newly acquired territories. She decided she should see for herself.

The rumors were true. Peasants were angry, there was resistance to her government, and a few royal heads might roll if the Empress found out how bad things were on the new Russian frontier.

Before her tour, a smart prince named Grigory Potemkin imported a few hundred well-fed Swedish and German farmers. He set them up with free land and new houses and cleaned up the few villages Catherine was likely to see up close. What he couldn't sanitize, he hid with fences and trees. Those projects became known as "Potemkin villages."

And they worked: Catherine did her big tour and went back to Moscow convinced things were just hunky-dory out there in her new territories.

The coal industry has its own version of a Potemkin village with the golf course in Floyd County. StoneCrest claims on its Web site to be 700 acres of successfully developed reclaimed land from mountaintop removal. I've seen pictures of it, and it's beautiful. It's also one of a kind.

Seven hundred acres is about 30.5 million square feet. A StoneCrest groundskeeper told me the intent was to put six inches of topsoil on the dead ground left by the coal operation. That's about 600,000 cubic yards of dirt; delivering it would require 38,500 double-axle, six-wheel dump trucks, more or less.

The coal industry thinks we're dumb enough to believe somebody would bring that much dirt to cover up the destruction of a mountain more than once.

The coal industry has other Potemkin villages. One is the federal prison in Martin County, built on a former mountaintop removal site. Defenders of the mining method are pleased to describe the prison as reclaimed land put to good use. The implication seems to be that Kentucky ought to be thinking of penitentiaries as a new growth industry, to fill up empty dead places where mountains once rose.

The implication seems to be that Kentucky ought to be thinking of penitentiaries as a new growth industry . . .

But the supporters don't talk much about how the project went $60 million over budget, due mostly to buildings settling. That devastated ground just can't support anything that heavy.

And they don't mention the nickname locals pinned to their new pen: "Sink Sink."

Empress Catherine probably got fooled because she wanted to believe everything was just fine with her empire. When that shrewd prince told her where to cast her eyes, she didn't bother to look past his prettified towns and imported farmers to see the disorder and rebellion beyond them.

The coal industry, determined to get to its version of black gold in the cheapest way possible, doesn't want the people of Kentucky to look beyond a golf course or a prison.

But I think they will.

Climb the mountains and get their good
tidings. Nature's peace will flow into you
as sunshine flows into trees. The winds
will blow their own freshness into you,
and the storms their energy, while cares
will drop off like autumn leaves.
—John Muir

Personal Statement

by Betty Woods

I grew up in Leatherwood, in a coal camp. It was a good life, everybody knew everybody. But it was hard because there were 10 of us in the family.

I was next to the oldest, so that was lot of responsibility. I had to get out and house-clean for fifty cents an hour. I would bring the money back home and contribute. It was a rough, hard life. But the families were close then.

I was involved with the union when the Southern Labor Union shut down at Blue Diamond Mine in 1992. We walked the picket line. At that time it was the women who got Glen Baker's trucks [scabs] stopped. He surrounded us and told us he was going to kick our #$%, but we didn't budge. There were scabs going across the picket line, so a bunch of us women figured we could at least help, so that's what we did.

I have lived here [in Ary] about four years. At first I didn't like it here. I hated the coal trucks, because they are so noisy and muddy. You couldn't sleep. But, you get used to it.

Gradually, I grew to like it here. You couldn't ask for better neighbors—they have been really good to us. And I have a daughter up here.

I don't like mountaintop removal.

We live near the second largest mountaintop removal job in the state. The way they do it around here is not right. There is always mud coming down into the creek, all the trees are gone, the watersheds are gone. Every time it rains, that creek will fill with mud. The creeks are filling with that sludge. We may wake up with whole hillsides on us.

The good Lord put the mountains there for a reason. He didn't put them there so we could cut them all down and make it wide open.

One time, I was laying in bed asleep and the most awful noise there ever was blasted me out of bed. It scared me to death. It was blasting from the mine.

You can see what it's done here all through my house. We have fixed the leak twice before—it keeps coming back. It has just ruined everything, and I can't afford to put it back. It has knocked pictures and everything else right off the walls.

Last year, I went through here and I put this stuff on the ceiling that covers the stains, and then I painted. Now the stains are back. We have been on top of the trailer working.

It's ridiculous. It's a shame that you can't have anything. We have called the company, and a man came out and looked and said we definitely had damage from blasting. That was two years ago and we haven't heard from him since.

It makes me mad because we can't afford to put it back. I think they, the company, ought to be held accountable for what they have done. The everyday person could not get away with this. But they got money and they can get away with it.

—1999

[M]ountaintop removal mining, by destroying homeplaces, is also destroying ancestral ground, sacred ground where generations after generations have lived, gone to church, married, made and birthed babies, taken family meals, slept in peace, died and been buried. The sanctity and sacredness of all life and the natural environment created by God should not be destroyed in the name of corporate profit.
—Catholic Committee of Appalachia, 1998

The Ghosts of Mountains

by Mary Ann Taylor-Hall

Last spring, I flew over an area of devastation where once a series of mountaintops had been. Of all the life that had been in that place, not one thing remained. It came to me, as I tried to take in the complex ramps and excavations and piles of rubble, that if you could see into the mind of some coal executive, far away in his glass office with his topographical maps, it would look like this site—a world where nothing has value or reality except what can be converted into cash.

All the earth's small intricacies, the wild phlox growing in the root of the great bur oak, the unexpected bank of violets, the deer and squirrels and little creeks and waterfalls, the snakes and salamanders and tree frogs, all things that make the earth real—gone to the great destroyer, the blaster, the digger, the dog pawing furiously for the buried bone through topsoil it took thousands of years to make, and then bedrock, too, the mountain and all its trees and soil pitched into the valley to get to the vein of coal.

I am grateful that I live in a place that has no coal under it. If what I saw below me befell a place that I loved, I would shortly die of grief and fury. I thought, looking down at the violence, "This is how the earth will look at the end. It will look like what we did to it."

Bill Caylor, the president of the Kentucky Coal Association, suggests that I am too easily outraged. I want to ask Mr. Caylor what it would take to outrage him. He thinks I have a bias against development. I think he has a bias against the Earth and all the creatures (including humans) trying to live here.

I wonder if he ever had a place he loved, a place he felt he knew through all its seasons, a place he wouldn't want destroyed. How would he feel if he watched its destruction and then was given, in its place, a Godforsaken featureless pile of packed-down gravel covered with coarse grass from hydro-seed, where no self-respecting tree could ever grow, and told to be grateful?

Now the forest canopy that softened the rain in its fall is gone, the small wet-weather streams that let the rain down slowly to the creeks are gone. The rain hits the earth sharply, takes a straight way down to the hollows where the people live. The silt-filled stream beds can't hold the runoff from even an inch of rain. The water rises quickly over the banks.

The children can't sleep at night, afraid of floods, afraid of speeding coal trucks. They can't drink the fouled water that comes through their plumbing. They can't play outside without being covered with coal dust.

Mr. Caylor magnanimously announces that the coal companies leave behind "valuable flat land for future generations." Maybe that's what he tells himself in order to get to sleep at night.

Greed is always with us, God knows, but greed that tries to pass itself off as public-spirited beneficence deserves to be hooted off the stage.

It is a miserable bequest. What our children will actually inherit from this insane practice is squandered wildness, polluted rivers, ruined roads, and the ghosts of mountains that were shelter for our spirits.

"I will lift up mine eyes unto the hills, from whence cometh my help," it says in the Psalms. It doesn't say, "unto the housing development, unto the golf course, unto the strip mall." These mountains never belonged to the coal companies, whatever the deed says. They didn't belong to me, either. They belonged to themselves.

To blast out a whole mountaintop and all its life! There is no forgiveness possible—nothing left to grant forgiveness, only the silence when the blasting finally stops, when the bulldozer shuts down.

P.S. Here's the saddest thing: Kentucky's energy needs could be derived four times over from renewable energy sources, without damage to the earth and the lives of the people who live near the coalfields.

Upheaval

fiction by Chris Holbrook

The cab of the truck feels hot already, and already Haskell can feel the film of coal and dirt gomming his skin. Clouds of dust rise high enough to pebble his windshield, so thick the roadway edge is barely seeable. He wonders who it is driving the spray truck and why they're not doing their job.

As he passes the raw coal bins, he meets George Turner flying toward him on his road grader, the machine bouncing so high on its tires that Haskell feels a gut-clench of fear. He thinks to himself, *That's too fast. He's going too fast.* He touches his own brakes as if to slow George Turner's grader and battles the urge to cut his wheels toward the road edge.

A shrill ringing begins deep in Haskell's left ear, like the whir of a worn bearing. He can see George Turner's face—the beard stubble on his jawline, the thumb-smudge of coal black beside his nose. It surprises Haskell how near George looks. He braces his arms for the hit, grits his teeth as the ringing grows more worrisome. Almost before he knows it then, and with no more a calamity of dust and motion than a hard wind might have caused, the grader is by and gone, not even near to tagging him, not really. The ringing in Haskell's left ear fades, becomes a little pin-prick of sound near the hinge of his jaw, so slight he can almost ignore it.

Lord have mercy, he tells himself. His headache has begun to throb more strongly, and he feels a weakness in his hands, as if he's gone too long without eating. He pops two aspirins and swigs from his water jug. It had been a solid chunk of ice when he'd taken it from his freezer that morning, but the ice is melted away now, the water almost warm. I'm as nervous as a cat, he tells himself. He drinks deeply, chasing the bitter taste of aspirin, letting the water fill and calm his stomach.

If I just get through this day, he tells himself. *If I just get through this day.* He presses the water jug to his forehead, feels its surface moisture seep into his temples. He sees the spray truck then, parked at the road edge next to the Peterbilt that Albert Long drives. It is Jim Stidham's boy, the one they call Tad, standing at the rear bumper, fighting to coil up a hose. As he passes nearer, Haskell sees the Peterbilt begin to back toward the rear bumper of the spray truck. He thinks, *Surely that boy's got sense enough to look around*

47

*hi*m. But Tad keeps on coiling the hose, neither moving nor looking up. Haskell thinks, *Surely he can hear the backing alarm.*

He tries to catch himself before he leans out the truck window and yells like a fool. He tries to tell himself that what he thinks he is about to see happen—Tad Stidham getting pinched between the truck bumpers, getting crushed, getting killed and mangled—is no more about to happen than he'd been about to collide with George Turner's grader. It's just him, him in his nervousness looking for the worst to happen.

He leans out the truck window then and bangs the door with the flat of his hand and yells, "Ho! Look out there! Ho!" Tad Stidham startles so suddenly that he almost trips himself on the hose. For a moment he looks wildly around. He looks at the Peterbilt that has stopped backing a good 10 yards away and is now pulling forward onto the haul road. He looks at Haskell going by in his big rock truck.

Haskell sees the expression of the boy's face change from startlement to anger, and he knows by that what the boy must be thinking—that Haskell's called warning has been just to scare him, has been just to make a joke of him again because he is low man on the totem.

The boy flings the hose to the ground and shouts something at Haskell. And though he can't hear the words, Haskell knows he's been cussed. He feels a touch of anger himself then. He doesn't cuss other men. For some reason he thinks of the look Dory had given him that morning, like she couldn't wait for him to be out of the house. He is angry still when he passes the walking dragline and turns his truck to get in line for reloading.

He watches the boom of the dragline swing out, a football field long. It is hard to think how big a piece of equipment a dragline really is, hard to see without some other smaller piece of machinery standing near for comparison.

He watches the bucket rake into the overburden. Tremors rise up through the tires and frame of his truck and up through his boot soles and legs like all the ground beneath and around him is being upheaved. It is hypnotizing. One haul, a hundred tons. He feels his mind ease some as he waits, knowing he has a good ten minutes or more of sitting idle. It occurs to him what he might have said to Dory. *I'll be gone here directly*, he might have said. *You'll be shed of me directly.*

The morning breeze has died, and though the air has grown less humid with the noontime heat, it feels even muggier now. Most of the men sit or half-lie on the ground or on the tailgates of their pickups as if to move even enough to eat is more of an effort than it is worth. Others pace on restless legs, and some stand and kneel at intervals, seeming to find no ease in either position. Once and again a man will speak to say how hot and miserable it is. They sit quiet otherwise and sullen in a way not common to any.

Haskell sits next to Joe Calhoun on the tailgate of Joe's pickup. He can still feel the motion of the rock truck in his legs and arms, and from time to time he reaches his hand to the pickup's bed wall, feeling the need to brace himself against something solid.

From the corner of his eye, he watches Joe Calhoun nurse his left hand. It is wrapped around from wrist to fingertips with a white handkerchief. Joe holds the hand away from himself as if neither to see it nor let it be seen. It bothers Haskell not knowing how he's come to harm himself—Joe the oldest man on the site, the least careless, the least likely to mistake himself around machinery. It is a battle not asking.

A moment later the old man absent-mindedly touches his bandaged hand to the tailgate of the pickup. He takes a sharp breath and brings the hand in close to his chest. He curses, so odd a thing for him that Haskell feels an urge to turn away.

For a while he watches Josh Owens and Bill Bates. They sit nearby, facing each other across a cable spool turned on its side for a card table. They focus on their play with heads lowered, hardly speaking but to bid or pass or call trumps, their behavior so out of keeping with their normal foolery that Haskell feels his own humor made bleaker by their close company.

When he looks again at Joe, the old man's face, dark as it is with sun and weather-burn, seems almost sickly. He is about to give over and ask Joe if he's all right, when a sudden, unexpected uproar commences among the card players. Haskell looks to see Bill Bates standing over Josh Owens, Bill flinging his cards in disgust upon their make-do table.

"That's enough," Bill is saying. "By God, that's enough of that." The sides of his neck and his ears have become suddenly blotched with red, and the skin of his face where it shows through the mask of coal-black has flushed red.

"Now, I don't mean nothing," Josh says, standing as well.

"You don't never mean nothing."

Haskell has half-risen to step between the two, when Joe Calhoun mutters something almost beneath hearing. Haskell leans toward the old man, straining to catch his words. "What?" he asks. "What is it?"

For a moment longer Joe sits completely still, his face pained and angry-looking. Suddenly then, he turns to Haskell and speaks in a strong, loud voice. "I said a man's got to watch. Watch himself and everbody around him. That's just the fact of the matter."

Joe stands and strides over to Bill and Josh. Without speaking, he gathers up their cards in his good hand and walks on. After a few paces more, he stops and turns, staring upward toward the job site.

For a while longer Josh and Bill stand facing each other. Slowly then their eyes begin to shy toward the ground, and Bill's face begins to cool until finally he looks more sheepish than angry.

For a while then Haskell sits studying the big insulated cable running from the generator house to the dragline. He tries to think how many volts it carries. 60,000? 80,000? It is a firing offense for a man just to walk near that cable. He sees Tad Stidham standing at the far end of the row of personal vehicles, and his own anger sparks again. *I don't cuss other me*n, he thinks.

Haskell wipes his shirt tail into his eyes, clearing sweat and fine grit from the corners. His vision clears for a moment, then becomes hazy again. He bites his sandwich, and even that seems mucked, the bread made soggy by the steamy air, the baloney flavored less by the taste of mustard than the scent of diesel fuel on his hands.

He thinks again of the ongoing dispute he's had that week with Dory. His hands almost shake as he brings their bickering to mind. *Why is it you all the time on me? All the time?* He pitches the rind of his sandwich toward the ditchline and takes a drink of water to rinse his mouth of the tainted aftertaste. He rises and walks over to where Joe Calhoun still stands, still staring toward the job site.

"Got to be on your guard," Joe says. "All the time."

"You'd think a man could get some little break," Haskell replies. "Some little peace of mind."

The other men begin to rouse themselves then. Those who are still pacing stop and stand staring a last few moments into the distance beyond the near ridges. On the leveled hillside above them, the dragline works on and the two payloaders work on to keep the outgoing coal trucks filled. As they start up the hill-path toward their machines, the men raise their eyes to watch the boom of the dragline swing about, the huge bucket darkening the earth askance of its path. Haskell feels a slight chill at the back of his neck as the ground where they walk becomes shadowed. His head begins to throb more strongly.

Half-an-hour before quitting time, Haskell hears it, a sound so faint as to be imagined, just barely out of kilter with the regular uproar of haulage—of backing alarms and rumbling wheels and buckets and blades and whirring auger bits—a lone, flat-sounding boom like far distant thunder, like something miles away, heavy and solid, coming hard to ground—followed by slow-growing quiet.

In ten minutes he is standing with the crowd of men on the incline above the overturned coal truck. He is watching the boy being lifted from the crushed-in cab.

Who is it?

That Sparks boy. That Pete.

Got away from him, huh?

Most of the coal load still lies within the side-turned truck bed, though spilled blocks strew the hillside from the road edge downward, marking the path of the wreck. Hydraulic fluid has begun to drip from a burst reservoir, a reddish stain forming as from a living wound within the litter of broken glass and metal and other odd rubbish.

Is he killed?

I don't believe he's killed.

He ain't killed?

No, he ain't killed, I don't believe.

Men crowd forward as the stretcher passes near. They reach, lifting, jostling, getting their hands in. The Sparks boy's teeth are gritted. His head and neck are held rigid in a brace. He must roll his eyes wildly to see about him. As he is lifted into the back of the ambulance, he raises an arm.

*He's a tough on*e, somebody calls out.

The rear doors of the ambulance close. Hands pat the sides of the vehicle in signal. Somebody yells, Ho! For a while then no one speaks. There is near silence across the job site. The dozers and rock trucks, the payloaders and road graders all are hushed. Even the dragline has yet to be restarted.

In the lack of machine noise, a soft roar can be heard, as of gathering wind. A bank of clouds crosses the sun, darkening and cooling. On the near ridge lines, the stands of fir and hickory, of beech and sassafras, begin to sway. Before long the wind is passing among the men, a cooling wash of air scented with coming rain. Then several voices start in together.

I seen a boy one time open a gas main with his dozer blade. Burned him and the dozer both up.

That day Arthur Sexton and Sonny Everidge hit one another head on.

You don't know what's like to happen.

They move about restlessly now, no longer tired out. They nudge one another with elbows, clap hands on one another's shoulders, jostling and play-horsing, a sudden wildness come upon them.

Sonny broke both his arms, his collar bone, three ribs, his ankle, fractured his skull. They say that Sexton never had a mark on him. Not a bruise.

You got to watch yourself.

Old boy I worked with at Delphi.

Their talk carries loudly into the stillness, their voices echoing strangely in and about the cutbanks and spill dumps and coal bins.

Told how a deer come leaping off a highwall.

Who was it to blame? That Everidge boy or that Sexton?

Watch yourself and everything around you.

Crushed in the cab of a man's pickup.

Joe Calhoun stands holding the wrist of his injured hand, staring into his bandaged palm as if to answer for himself some puzzlement. From time to time he looks up and shakes his head and laughs aloud.

It's just something that happened is all.

You got to watch. All the time.

I don't remember if he was supposed to been in it or not.

One by one then they hush, their high-sounding laughter falling still, their unruly moods dampening. Finally they all stand quiet again in consideration of the wreckage. They stand as if praying, as if thinking together a single thought.

Bill Bates and Josh Owens are first to quit the assembly. They head off together down the hill path. But for Bill's height and Josh's width, they could have passed one for the other, their clothes covered from cap to boot-toe with grease and oil, their faces and hands blacked with coal dust. They walk near side-by-side, conversing in friendly-seeming terms.

Slowly the work noises start up again. As he leaves the job site with the other men, Haskell feels his own day-long pall of nervousness and worry begin to lift, his mood turned about amid passing talk of knife brands, of heat and dust, of weekends planned fishing.

He thinks now of rest and stillness. *Maybe set on the back porch awhile.* Maybe set on the back porch with Robbie, awhile. He smiles, thinking of his son. His headache has almost completely eased. *Maybe set on the back porch with Robbie and Dory both. Set there till dark. Set and rest.* He thinks of a promise Dory has made him, of hamsteaks and turnip greens some night. He feels overcome by a moment's strong sentiment, that followed by a deep, almost painful hunger. *Why, she does as best she can,* he thinks.

For a while after he gets home, Haskell rests on the top step of the back porch and views his garden. It's as fair a garden, he knows, as a man could hope for, what little time he's had to give to it. The pea pods hang full, their vines thickly twined along two rows of wire fencing. (They will want picking in another day, while the hulls are still yet tender.) The stalks of corn that have come up are shoulder-high and beginning to tassel, though there are spaces left in the rows where the crows and ground squirrels had got at the seed and where he'd not had time to replant before the season got too late. The beans are full of bloom. The Irish potato patch looks blighted and eat-up with bugs, though; and the row paths are greened over with weeds so that the garden is all over snaky-looking.

For a moment Haskell rebukes himself for having planted the patch at all, for having made himself slave to so much chore-work for such a little bounty. It will need hoeing and dusting and fertilizing still more, and watering if it doesn't rain. *Work on top of work*, he grumbles. And yet he feels no real downheartedness.

He throws up his hands. Bath and supper. Watch a little TV. Go to bed. His mood uplifts at this resolve. He reaches to undo his boot strings. It is a struggle, so tired and stiff have his fingers become.

He pauses to watch the shadowed passage of clouds across his pea patch, his corn; the westering sun showing itself in slantwise fashion from just above the ridgeline, brightening vines and leaf blades. For a moment he feels like he might fall asleep right where he sits.

There is a loud slap of the screen door flying open and shut behind him. Haskell startles. He has only half-risen when Robby comes leaping upon him, arms flailing wildly, screaming, "Hey, Daddy. Hey, Daddy."

Haskell nearly loses his balance catching his son, gets the crown of the boy's head beneath his chin, a flung fist against his ear. He fends him off, stands him by force at arm's length upon the porch; the boy reaching with his hands still, chattering wildly still about twenty-eleven different things—some animal he has seen, some cartoon show, some wrong done him by his buddies at school.

"Robbie," Haskell says, and when the boy still does not calm, again, more loudly than he intends, "Robbie!"

The boy quiets then, as suddenly as if he has been slapped, and Haskell sees it coming. He looks toward the screen door, is at once relieved not to see Dora standing there. As quick as he can, he gathers the boy in his arms and walks with him down into the yard. "Don't cry," he says. "Please, don't cry." But the boy is beyond shushing. He begins to blubber wildly, struggling against being held now, struggling to be let go, to run with hurt feelings to Mommy.

Haskell holds the boy tightly. "Please don't cry," he says. "Daddy loves you. Please don't cry. Daddy loves you." After several minutes of coddling, the boy calms enough that Haskell can chance to tickle him under his arms. "Gee, I'd be ashamed," he says, "big boy like you." Haskell jostles his son up and down, gets a hesitant, sullen smile, then swings him in the air in a mimic of flight, finally coaxing the boy back around until they are friends again.

She is there when he carries Robbie through the back door into the kitchen, there where she always is when he comes in of an evening, still wearing the red Food Town smock that is the first and last garment he sees her in each day. She reaches to take the thermos and bucket from Haskell even as she sizes the situation, of Robbie slumped and clinging, red-faced upon Haskell's shoulder.

The look she gives him, Haskell knows he has coming, still yet it pisses him off. I've not been home five minutes, he wants to say. *Not even in the door, hardly. Not even got my shoes off, hardly.* But he gets no chance to speak.

Robbie starts in again, taking Haskell's face in his hands, forcefully turning it to him, his talk a breathless gabble of noise out of which mess Haskell hears plainly no more than, Dad. Hey, Dad. Dad. Haskell stares Dory down, his own look as purposeful as he can make it. *Can you do something? Now? Can you?*

She tilts her head back, shaking it slowly, rolling her eyes toward the far wall as she reaches for Robbie. Haskell passes through the kitchen with Robbie still clamoring after him. He pauses just long enough in the hallway to watch Dory kneel before Robbie, to watch her place her hands on his shoulders and speak to him in a stern almost harsh manner that still yet settles the boy enough to hush him. *I've not been home five minutes,* he wants to say.

He runs water into the bathtub. The water has a sulfur-like smell, but it's fairly clear. It warms up but will not get all the way hot. *Tank needs its filter changed out again,* he thinks. Chill bumps rise upon his shoulders as he settles himself in. He feels the sting of bug bites on his legs, of cuts and scrapes on his hands and arms and sunburn on his neck, though in a little while the water soothes him almost to numbness.

Haskell feels some better after his bath, more like himself, he thinks. Before supper, he lounges with Robbie on the living room floor, playing Crazy Eights. Robbie sits right up against Haskell, and for a long while Haskell is unable to move even enough to stretch his legs without drawing a worried look from the boy. He is relieved when Dory finally calls them to the table for supper.

Haskell is disappointed to see fried baloney on his plate instead of hamsteaks. Even so, he feels a powerful sense of well-being—the smell of the warm food (meager though it is), the knowledge of himself, his wife and child all together in the close kitchen (safe from harm at the day's end). He has a strong impulse to say a blessing, is on the verge of doing so when he feels an even stronger foreboding at saying or doing anything that might disrupt what he sees as his family's shared moment of contentedness.

They all sit quiet through their meal together, Dory hardly raising her eyes from her plate except to glance toward the window or the wall clock. Robbie has calmed some since their card game together, though Haskell is still half-afraid to move lest he give the boy some unintended upset. He begins to wish more and more strongly for conversation, a joke, anything to cover a little the irksome sound of forks scraping against plates, of chewing and swallowing, but there is nothing he himself can think to say now.

By the time Dory rises to clear the table, Haskell has begun to feel the first mild throb of his daylong headache coming back on him. Listening to

Dory clang away at the plates and bowls, it occurs to him to ask if that noise is necessary, to ask if that noise is in some way for his benefit.

"Set down here and rest a minute," he says instead. She continues to move as if she has not heard him, and for some reason (he is not sure why) he feels it foolish to repeat himself. He feels a growing urge to strike something—the wall, the table. He can see himself upending the table, kicking his chair across the room. Something, anything, just to make known what he thinks of the situation.

He feels a tug on his arm and hears vaguely the sound of Robbie's voice saying something again and again. He blinks his eyes open, though he'd not been aware even that he'd had them closed. Dory is sitting across from him at the table again, watching him.

He speaks without knowing he is about to. "You'd think a man could have just a little peace of mind when he comes home of an evening," he says. "You'd think a man could look forward to a little rest in his own home."

"You can rest," she says.

"Ain't no peace of mind around this place."

"You can rest," she says. "Nobody's stopping you."

Haskell says nothing in response. In his oncoming gloom he has begun to think again about Pete Sparks wrecking his coal truck, about the sound it had made going over, about the way the boy had looked being pulled from his truck cab, him shaking so hard from shock you could hear his teeth clatter as far up as the roadway. He thinks again of his own near mishap, what he believes was a near mishap, with George Turner's grader; and he feels again the urge to touch his hand against something solid, to steady himself against the fitful dizziness and upset he feels come upon him again, against the tiredness.

He looks at Robbie. The boy is staring at the bare tabletop, his face stiff, a flush of redness on his cheek and neck. Dory places her hand upon Robbie's shoulder, and he seems at once to calm. Without speaking again, Haskell rises and walks out of the kitchen.

He stands on the back porch a long while, feeling the comfort of the moist night air. When he finally feels some at ease again, he turns and looks through the screen door into the kitchen. Dory and Robbie are still seated at the table, their heads close, speaking so softly together that Haskell can barely hear the sound of their voices.

Dory's hand is still on Robbie's shoulder. Haskell notices the small bandage on her finger. He notices how worn and dirty it seems, as if it has been there for days. He notices too the dark streaks of grime on her smock, the spots of grease, the blue pricing ink staining her fingers. He sees how her shoulders slump in a way that he knows is not tiredness only. It occurs to him that she would have gotten home hardly before he did.

He blinks and shakes his head as if to rid himself of some unseemly spectacle, of some bad odor or vexatious thought. *I don't see what she's got to complain about,* he thinks. He'd like to tell her how bad his legs hurt, how bad his back hurts. He'd like to tell her about Pete Sparks wrecking his truck. He'd like to tell her what a day he's had. He reaches his hand to the door, pauses, then he walks off the porch into the dark of the yard.

Haskell watches his truck's rear end in the side mirror, lining it with the berm as he backs toward the highwall edge. His neck feels tight, and already there is a dull pressure in his temples that throbs with each shriek of the truck's backing alarm. He feels in his shirt pocket for the aspirin tin, then suddenly the muscles in his back and legs and arms all clench at once and he hits his service brakes. He leans out the window and looks hard at the ground before the berm.

The truck's engine throbs through his chest, and for a moment it is as if his heartbeat rises and falls with the idle speed. He tastes diesel at the back of his throat, feels the sting of it high in his nostrils. His head swims like he is drunk. He fumbles for the seatbelt catch, and then he realizes if it was going to go it would have gone already. He sucks deep breaths. It was not the ground giving way, he'd seen. It was heat shimmers. Or it was the shadow of a cloud passing. Or it was light on his mirror.

For a while he watches Joe Calhoun working his D-9 on the adjacent hill-seam, the dozer's blade cutting into the overburden, loosing boulders and small trees toward the valley floor. It seems a marvel almost, the way the huge dozer clings to the contour of the hillside, the way the tracks sidle and shift on the near-vertical incline.

He watches as Joe Calhoun goes about leveling a large beech, first ditching the ground on the downslope and then above. In a short time, the tree begins to topple of its own weight, its branches catching and snapping against the still-standing timber, its roots tearing slowly free of the ground. Joe Calhoun moves the dozer to and fro, nudging with the blade in a gentle-seeming way.

Haskell feels himself begin to calm. He has run a dozer himself. It is as familiar as any piece of machinery on the site, but he watches it now like a man seeing something he never has seen before. He feels strangely like he is about to see something or know something he never has, that all he has to do is sit still long enough and watch close enough and it will come to him.

"He knows what that machine can do," he thinks. "He knows what he can do on it." But as the beech begins to skid down the hill slope, its broken limbs shining whitely in the bright sun, clods of black dirt dripping from its tangle of upturned roots, he feels again the sensation of loose soil sliding beneath his wheels. He presses his foot even harder onto the brake pedal.

From the corner of his eye, he sees Ray Sturgill sitting in his truck, waiting his time to dump. Haskell wonders how long a while he's been sitting idle just watching another man work, if it's long enough for Ray to have thought something.

He lets off his service brakes and continues backing until he feels his wheels touch the berm. Then he puts the transmission into neutral, sets the park brake, and pulls the dump lever. At the same time, he guns the engine. Dust rises with the clamor of falling material from his truck bed. He can feel the truck's back end jarring, and for a moment he feels dizzy again. He clenches his hands tight on the steering wheel and raises his foot above the brake pedal. But then the bed empties out and the rear end is still again and he is not slipping off the highwall edge but sitting stable.

He lowers the bed, puts the truck in gear, and lets off the park brake. The adrenaline fades out of him as he pulls back onto the haul road. He begins to worry then that he's forgotten something important or overlooked something important he should have seen. He runs through his morning safety checks— belts, brake linings and pads, wheel cylinders, hydraulic lines, air tanks. There is nothing he can think of that he's missed.

Two Replies

The following are replies to a newspaper column in which Steven Gardner supported mountaintop removal and questioned the motives of the Kentucky authors who oppose that mining method.

Mining Destroys Land and Spirit

by Silas House

I take issue with everything Steven Gardner said in his commentary about mountaintop removal, but one error that must be corrected is his claim that the authors' tour of mountaintop-removal sites was paid for with public funds. We paid our own way. We paid for our own food, gas, lodging, and whatever else it took to make the tour work because we believed in what we were doing.

Another point that must be clarified is that we're not trying to turn fiction into fact. I have merely reported what I've seen with my own eyes. I can't comprehend anyone thinking that any kind of erosion is a good thing, much less "accelerated erosion," as Gardner calls mountaintop removal.

The only thing that the coal companies and their supporters want to talk about is reclaimed land. That's their only playing card. Even if land were properly reclaimed (and very often it's not), why is it that they never take personal responsibility?

Gardner claims that "when mistakes are made, people or companies should be accountable." But all one has to do is check Appalachian history (the Buffalo Creek disaster, the Harlan County strikes, the Martin County sludge spill, etc.) to see that the companies have not been accountable. When they make a mistake that costs taxpayers millions of dollars, they chalk it up as an "act of God."

What Gardner failed to reveal in his commentary is that he is closely connected to the mining industry. A quick Google search yielded five pages of links about Gardner's intimate ties to the coal companies (and several similar commentaries, all written before the authors' tour).

As president of a large engineering firm that makes money from the coal industry, Gardner is only protecting his assets. And that's fine and respectable. However, it is not fine or respectable to say that members of the

59

authors' tour are writing fiction about the coal industry. It's the oldest insult in the world to accuse writers of being liars, and I am sick of it. We're not making things up, and most people know this.

Like many of the authors involved, I am striving to be better educated about the practice of mountaintop removal. Gardner seems to claim he knows everything there is to know about the subject, but I don't. And I've never acted as if I did. But no matter how many

I'm going up on the mountain,
and I ain't comin' down till morning.
I'm going up on the mountain,
and I ain't comin' down in chains.
—Traditional

statistics are thrown on the table before me, I can't deny what I see every time I drive across Eastern Kentucky. I can't turn a silent ear to the stories people tell me about living in the shadow of mountaintop removal. Unlike Gardner, I've never played golf, but I have lived in Eastern Kentucky and experienced the ill effects of mountaintop removal.

I'm not completely against coal mining. Without it, my family would probably still be poor. But it can be done in a more responsible manner. That's really all the authors are demanding.

I don't care how someone might benefit from this leveled land 50 years from now. I think only about the parents who can't afford to buy cases of bottled water or go golfing because they're too busy drilling a new well every other month to replace the ones collapsed because of blasting, or getting their cars repaired because of roads destroyed by overloaded coal trucks.

Even Gardner says that "mining is inherently destructive." What he fails to state is that we are talking about more than destroying land, which is bad enough in itself. We're also talking about how the spirit of the people is being damaged. Those people know that it's wrong to make a profit off others' suffering, just as it is wrong to make a profit by destroying the land. And that's what's happening in Eastern Kentucky.

The Facts Aren't Pretty

By Erik Reece

Reading J. Steven Gardner's dangerous paean to mountaintop-removal mining is no way to start the day. I would hesitate to return volley in this "he said/she said" debate over strip mining in Kentucky if the stakes were not so high.

Since I wrote an article about this poisonous form of mining for *Harper's* magazine and since Wendell Berry organized the authors' tour of mountaintop removal, several representatives of the coal industry have accused us of making "emotional claims."

Gardner accuses me of writing "fiction" (a charge that will surprise the professional fact-checkers and lawyers at *Harper's*). But since Gardner wants only the facts, let me supply a few.

In the last three years, 14 people died in West Virginia because walls of mud and water roared down from mountaintop-removal mining sites. That's a fact.

In the last five years, more than 50 people have been killed in Kentucky by illegally overweight coal trucks. That's a fact.

In Virginia, a boulder rolled 200 feet down from a mountaintop removal site and crushed to death a sleeping three-year-old, Jeremy Davidson. That's a fact; just ask his parents.

Gardner writes that mountaintop mining's critics overlook the reclamation of mine sites. But the truth about reclamation in Kentucky is that it is usually done poorly or not at all. To replace the most diverse ecosystem in North America with a monoculture of invasive, often exotic grasses is cheap, easy, and irresponsible.

But the real story is that there exists so much abandoned mine land in Appalachia that the Department of Interior doesn't even try to account for it all. Coal companies have to put up bond money before they receive permits to mine. If a company does not stick around to do any reclamation, the bond money is meant to cover the cost, but it is never enough to do so.

What do the coal operators do? Too many simply declare bankruptcy, leave the mine site as a toxic hazard, and start up a new company somewhere else. That is why there are so many gray, lifeless plateaus spread across Eastern Kentucky, sites that may never be reclaimed.

There are certainly good people trying to mine coal with the least damage to people and property. But a corporation has one goal: the bottom line.

Don Blankenship, chief executive officer of Massey Energy, knows that his overloaded coal trucks are killing people. So why won't he rein in his drivers? Massey Energy knew that its slurry pond in Martin County would eventually give way. Yet no repairs were made, and the result was the 300,000,000-gallon slurry disaster of 2000.

Finally, as a teacher who regularly takes writing students to the University of Kentucky's Robinson Forest, I was most disturbed by Gardner's suggestion that UK "should do the right thing" and mine more of the forest.

I want my students to understand that the natural world—Robinson Forest in particular—holds higher values than that, including the spiritual, the aesthetic, the ecological, the sustainable, the redemptive. Those values may not be "facts," but they are the basis of a humane, democratic culture.

Why I'm All For
Mountaintop Removal

by Jeff Worley

I'm from Kansas and in Kansas
the earth is flat. If something ain't flat,
it ain't right. You walk anywhere,
you go right from A to B. And that's that.
When I was a kid in Curly's Barbershop,
Saturday mornings, he didn't even ask.
It was flattop, flattop, flattop, a line of us
leaving the shop as God intended.
These mountains in Appalachia?
They need to be taught a thing or two
about plain. About who's boss. If God
didn't want these mountaintops
sliced off, why'd he invent the dozer
and dragline? Ask yourself that.
The earth is flat. We all know that.
No getting around it.

Sayings of the Apalachees:
From a Thousand-and-one Original Proverbs

by Steven R. Cope

#354 The earth will rebel; it has already objected.

#451 The polluter of a stream should be made to swim in his own wee-wee.

#529 There's nothing like a mountain to teach a man right where to stand.

#626 Animals have given up wondering what on earth we are doing.

#642 The tree knows very well when the fruit is picked; and the weed with resignation gives way at the root.

#663 There is a sort of humanoid that would destroy a mountain for a nugget, drain the sea for a pearl, annihilate a species for a tusk.

#685 Men who cut tops off of mountains need to be baby-sat.

#763 Do not trade a mountain for an alley, no matter how many cats come with it.

#814 A state that cannot keep its river clean should not be permitted a river.

#972 When the old are gone, the young will sell.

#974 The path up the mountain is the mountain.

The Impact of Mountaintop Removal

based on a Federal Environmental
Impact Statement and research by
Kentuckians For The Commonwealth

Mountaintop removal—and the resulting damage to surrounding forests, streams, and valleys—has already had a profoundly destructive effect. If nothing is done to stop mountaintop removal, the future will be even worse.

Loss of Jobs

▶ Mountaintop removal contributes significantly to the loss of mining jobs. Many fewer miners are needed for this type of mining than for traditional underground mining.

▶ Strip mines produced an average of 32% more coal per worker than did underground mines in 1995-1999. Thus, production levels could be maintained with a third fewer mine workers.

▶ Over the last 23 years, coal employment has fallen sharply. In 1979, there were 35,902 mining jobs in Eastern Kentucky. By 2003, this had fallen to 13,036 jobs—a 60% decline. In other words, more than half of the jobs were wiped out.

Impact on Land and Forests

▶ 61,880 acres of valleys have been filled with rubble from mining operations. All life forms in those valleys has been either displaced or destroyed.

▶ 2,200 additional square miles of Kentucky's forests are likely to be demolished by large-scale mining in the next seven years. An area this size would encompass the following counties: Floyd, Knott, Leslie, Letcher, Perry, and most of Harlan. The wildlife habitat and recreational value—as well as the lumber value—of those forests will be forever diminished if not obliterated.

▶ The federal Environmental Impact Statement says there is "no evidence that native hardwood forests...will eventually re-colonize mountaintop mine sites using current reclamation methods."

▶ The same Statement says that large-scale surface mining "will result in the conversion of large portions of one of the most heavily forested areas of the country, also considered one of the most biologically diverse, to grassland habitat." What in the world are we doing to ourselves and to our land?

Impact on Water

▶ 724 miles of streams across central Appalachia were buried by valley fills between 1985 and 2001. Those streams were wiped out of existence—they're gone. And mining companies have permits to bury many more miles of streams.

▶ 422 miles of streams were buried in Kentucky alone.

▶ An additional 1,200 miles of streams were damaged by valley fills.

▶ Unless restrictions are passed, another 1,000 miles of streams will be buried in the next 10 years.

▶ All life forms in these valleys have been—or will be—displaced or killed by these practices. And life forms downstream are being harmed or killed. The impact on our overall environment is incalculable. We can destroy life forms, but we cannot recreate them. If we learn that we needed the creatures we wiped out, it will be too late.

▶ No scientific basis could be established for environmentally "acceptable" damage to our streams and rivers. The only safe conclusion is that none of this damage is acceptable.

▶ The federal Environmental Impact Statement says it is "difficult if not impossible to reconstruct free-flowing streams on or adjacent to mined sites." Destroyed streams are probably gone forever.

▶ Stream monitoring shows that water downstream from mountaintop-removal operations is toxic. We are poisoning wildife, our environement, and ourselves.

Is the coal we get from mountaintop removal worth the price we are paying? How can it possibly be worth it?

For the first time in history, we have the power to destroy our planet. And that is what we are doing.

From the rise, he looks out over his place. This is it. This is everything there is in the world—it contains everything there is to know or possess, yet everywhere people are knocking their brains out trying to find something different, something better . . . but what he has here is the main thing there is—just the way things grow and die, the way the sun comes up and goes down every day. These are the facts of life. They are so simple, they are almost impossible to grasp.

—Bobbie Ann Mason

PART TWO

The Sum Result of Speculation

by Maurice Manning

Three long-legged paces is a rod on flat
ground; four is a Kentucky rod. One acre
is one hundred sixty square rods. Marking
off thousand-acre parcels is a lot of paces,
a lot of steps to count in your head,
especially if you have a bad tooth tolling
every step you take. Sometimes I would
take thirty-five paces in one direction,
then forty in another and say I had an acre;
sometimes I zigzagged through this rolling
country and guessed. One time I dug
a rotten tooth out of my jaw with a rifle flint
so I wouldn't lose count. One hundred twelve
paces is a mile. Two thousand acres is three
square miles. I don't know how many
lonely paces my feet have trod. I don't know
how many trees I've marked with numbers,
or how often I've been generous, or if
I should be faulted. I'm sad to say,
walking this country for money only
brought me loss; but I never once got lost.

I. Appalachia Meets the Outside World

by Loyal Jones

The values in East Kentucky and the rest of Appalachia are old ones, and they persist. Some say that is because we were isolated, because mainstream culture had passed us by and we were not a part of the economic system. But how could this be:

When even back as far as the Revolution, armies (which we also joined) marched through and fought in the region?

When in more recent wars, we were sent all over the world in higher percentages than the rest of the nation?

When mainline churches sent more missionaries to save us (who were already Christian) than to any other places in the world, except perhaps Africa and China?

When financiers and entrepreneurs from the major corporations bought up the timber, coal, gas, and other valuable property, established sawmills, mines, smelters, and hired thousands of us; and Appalachian timber, stone, and bricks were part of those New York and Newport mansions and corporate offices all over the country?

When hundreds of local color writers, journalists, folklorists, musicologists, anthropologists, sociologists, historians, and the like, roamed the mountains seeking material for a plethora of books, articles, films, dissertations, and theses?

When the Tennessee Valley Authority, the Corps of Engineers, public power cooperatives, and private power conglomerates built dams over much of the region for flood control or power generation and added thermal plants to utilize Appalachian coal?

When government workers swarmed through the region in the Great Depression and the War on Poverty fervently determined to "uplift" us?

When we embraced radio and television and other electronic inventions as sources of information and entertainment?

When three or four million families from Appalachia joined the migrant streams after farming and timbering and mining could no long employ them?

When millions of tourists came to view our natural beauty and to observe our alleged quaintness?

When drug dealers also invaded the mountains, finding customers among the unemployed, the disabled, and the hopeless in a worsening economy ?

Isolated indeed! But there is no doubt that every wave of visitors did change us and our culture despite our resistance.

The Bigger the Letter,
the More It Weighs

by Maurice Manning

If you like a thing right good,
it means you really like it a lot,
in fact, you wouldn't mind more of it.
Take mashed potatoes: you might say,
them's right good mashed potatoes, yes'm,
and heap your plate with seconds or thirds.
You might like picking beans right good,
how hot your back can be from bending
over the rows, and how you don't know
your back is heating up until
you stand up straight again and feel it;
so even pain can be right good
if you know it has a reason. Reason
is almost always pretty good,
at least the ones that, like it or no,
are true, and the reasons that aren't are the ones
you best stay shy of, friend. Of course,
there's reading through the older stuff,
where words like Tumult, Craft, Delight,
and Suffering are ennobled with what
they deserve, a capital—Tumult!
with a king-sized T! Delight! by gum,
with a full-grown D! The upper case
is like a hillside on the page,
a promontory, or a tree
grown taller than the rest. You like

that kind of landscape, alright, you like it
right good, the beauty, the treachery;
it's what your life has been and all
it's ever going to be: a rise,
a fall, and after that, you have
good reason to want another rise.
This is the land that bore you and made
you be, by Gum, by Sassafras,
by even Blasted Ash, by all
the Steeps and Vales, the Sunken down
the Lifted up, and the Waters run
between the two like common Tears,
though we both know there's nothing, no,
that's common here, and if you could give
the upper case to everything
that matters in your hilly world,
from the tallest Tree to the smallest Fry,
you'd have a sho'nough heavy Book,
alright, yes sir, no end in sight.

II. Mountain Humor

by Loyal Jones

Strangers have kept coming into the mountains, and while the stereotype said that we were a suspicious lot, one of our main values was hospitality, and so we invited them to come in and visit awhile. We found out that we were different from them. They said so. We didn't talk the way they did. We answered questions with stories. Our conversations were circular. It was hard to tell when we were putting them on, making some fun out of those who were making fun of us.

A missionary asked this old man if he had lived here all of his life, and he said "Not yet." Then the evangelist got down to his mission and asked, "Brother, are you lost?"

"No, I've lived here all my life."

"I mean, have you found Jesus?"

"Well, you know, I live so far back up this holler that I don't get much news, and I hadn't heard that he was lost. The Bible says he's up yonder (pointing upward) until the he comes again. You've heard of the Second Coming?"

"Look, I'm trying to find out if you're ready for the Judgment Day."

"When is it?"

"We don't know these things. It may be next week, or it may be next year."

"Well, when you find out, you let me know, 'cause the old woman will want to go both days."

Mountaineer humor is laced with irony.

A farmer over in the Breaks of the Sandy was dispiritedly hoeing his corn in a field with more rocks than soil when a preacher rode by and yelled to him, "Where are you going to spend eternity, in Heaven or Hell?" The farmer laconically replied, "Well, either one would be better than what I've got here."

Another farmer was asked how for he had to go to work. He said, "I just wake up, and I'm surrounded by it."

A lot of tourists came by, and they liked to talk to us natives and size us up so they could tell the folks back home about us.

One stopped his big Lincoln close to the fence where a farmer was working, and asked, "How far is it to Pikeville?"

The farmer said, "I don't rightly know."

"Then how far is it to Lexington?"

"I don't know that either."

"You don't know much, do you?"

"Maybe. But I ain't the one that's lost."

How it began. . . .

Why Coal Companies Favor Mountaintop Removal

by Maurice Manning

They want to do away with shadows, which
has always been their scheme, because they think
if everything is brought into the light
then all the other things they want to do
will be easier. But I would say we need
the shadows and the things which cast them out
across our naked plains. A shadow hides
the shame of knowing we're not efficient all
the time. And a shade is not a machine; it won't
break down or need to be designed again,
and that's reliable: let's say we need
to count on something being there for us.
I say the haunted look, the look that keeps
some darkness in it, is the sovereign face,
the face which always finds a way to find us
and reminds us we require some modesty.
My people, darker than God sometimes, I see you
in the shade of mountains, and if I burned a piece
of coal to see you better, I would burn
the darkness from it, but the darkness would
return behind the nearest object; it starts
on the other side of where the light runs out,
but let's agree that is the harder side,
where darkness makes another kind of light.

III. Mountain Music

by Loyal Jones

The traditional culture of the mountains was enriched by things from the far past, that is before all our interaction with the rest of the world. We told Old World folktales about such characters as Jack, Nippy, Mutsmeg, and Ashpet, that started out with the ancient words, "Once Upon a Time. . ." or just "One Time. . ." with a pause. And then came the long tale that had been carried over the waters in the memory of a forebear. Pioneer Kentucky folklorist Leonard Roberts published *The Sang Branch Settlers: Folksongs and Tales of a Kentucky Mountain Family* about the Couch Family of Leslie and Harlan counties, who gave Roberts 100 folk songs and ballads, and 164 folktales and riddles. The Couches were no strangers to industrial society. Jim and Dave Couch had both been coal miners, and Jim had lost a leg in the mines. Tim Couch, of football fame, is a relation.

Many talented musicians from the mountains found their way to places like Chicago, Nashville, Wheeling, Knoxville, or Springfield to make a pretty good living on recordings, radio, or television. There are still a lot of people in the mountains who sing the fine old lyric love songs and ballads about lords and ladies, murder and other tragedies, with some pretty fine lines and imagery.

Come all ye fair and tender ladies,
Take warning how you court young men.
They're like a star of a summer's morning.
First they'll appear, and then they're gone. . . .

Oh, love is handsome and love is charming,
And love's a jewel when it is new.
But when it's old, it groweth colder,
And fades away like the morning dew. . . .

My love is like a budding rose,
That blooms in the month of June.
Her voice is like an instrument,
That has been lately tuned. . . .

79

'Twas in the merry month of May
The green buds they were swellin',
Sweet William on his deathbed lay,
For the love of Barbry Allen. . . .

Oh, don't you remember, a long time ago,
Two little babes, their names I don't know,
Were stolen away on a bright summer's day,
And left in the woods, I've heard people say. . . .

It was late at night when the lord came home,
Inquiring for his lady-o.
The answer that came back to him:
"She's gone with the Black Jack Davy-o
She's gone with the Black Jack Davy-o". . . .

Down in some low green valley
Where the flowers fade and bloom,
It's there that Pearl Bryan
Lies mouldering in her tomb.
She died not broken-hearted,
Nor from disease she fell.
One moment's parting took her
From the ones she loved so well....

Personal Statement

by Pauline Stacy

In Lick Branch, there used to be homes and a school. The coal company moved people out and filled in where people lived. There used to be a lot of families, but there's no one that lives up Lick Branch any more. They also moved a cemetery out.

They took the tops off the mountains and filled it all in. That's when the creek started changing. I don't know what they did to it, but it's drying up now. It was a year-round stream. Now it only runs when it rains hard, and then it's more forceful, like a gully-washer. It comes off the hill a lot faster. Then when it stops raining, the creek dries right back up.

When it does run, the creek is from bank to bank, and not very deep. It's filled in with silt from the valley fill.

This is not just here. It's just about everywhere they've mined in Eastern Kentucky.

Some people say that because coal provides jobs in Eastern Kentucky, we shouldn't talk about the harm it does to communities: flooding, blasting damage, bad wells, and dust and mud from dangerous coal trucks. These things are against the law, but state and federal officials don't enforce the laws and people living in the coalfields pay the price.

Coal companies don't have to do these things to mine coal. They can mine coal and follow the law. But they choose to cut corners because they are greedy.

The coal lobby even goes as far as saying that coal companies should be able to break the law because coal gives some people jobs. But when companies break the law, it hurts everyone. Blasting destroys coal miners' homes, too. Mountaintop-removal mines, which are illegal in almost every case, provide few jobs compared to deep mines, but their blasting hurts homes around them for miles. The land and water our grandchildren will need is destroyed.

Companies that treat the environment terribly and get away with it often turn around and treat their workers poorly, too. Coal miners aren't as well off as they used to be. We have fewer and fewer union mines every year. The poor quality of life that careless mining creates—bad water, destroyed mountains, dangerous and dusty roads—keeps people from wanting to live in or even visit Eastern Kentucky.

The coal lobby will tell you America needs cheap coal to keep power bills low. But for us, it's not cheap energy. Bad water, blasted homes, and floods are expensive. We pay and pay.

1960, Deep Mine Workers

I live near a Horizon/Starfire mountaintop removal site in Ary [Perry County]. The devastation of mountaintop removal is enormous for residents across the state, but particularly for those closest to the strip jobs. The blasting destroys our foundations, our homes, our water wells, and our mountains.

The silt-and-sludge ponds that the coal industry puts in above our homes ruin our state's waterways and cause flooding and landslides.

Sediment from these mine sites fills up our creek beds and negates the flow of water, causing property damage and flooding for citizens in low-lying areas. The coal companies say that silt-and-sediment ponds are safe.

If they are as safe as the companies say, why do emergency spillways not hold? An emergency spillway on a Starfire mine site has given way, destroying my retaining wall, and eating away at my property.

Sediment continues to fill in our creek beds. What used to be a small branch and headwater stream is now much like a river when it rains. This pattern can be seen in communities across **I am here to tell you that these floods are not an act of God.** Eastern Kentucky and throughout the coalfields. Coal companies, regulatory agencies, and our media pretend that this devastation is an "act of God."

I am here to tell you that these floods are not an act of God. This destruction is an act of an outlaw coal industry. I have lived in the same location for thirty years and this never happened before. It did not happen until the coal company stripped our mountains, filled our headwater streams, and allowed their silt ponds to leak, break, and rupture.

Unfortunately, our regulatory agencies are not doing their job. Instead of protecting the citizens of this state, they are protecting the coal industry. Folks here are afraid to stand up to the coal companies because they say that mining brings jobs to Kentucky. Mountaintop removal and strip mining provide fewer than 12,000 jobs in this state.

The companies are using explosives and modern technology to remove the coal faster and cheaper with fewer employees. The coal industry is actually removing more tons of coal per employee then they did ten years ago, causing job loss in the region.

The people of this state deserve better jobs, better water, and better enforcement of our state and national laws!

They that have power to hurt and will do none . . .
They rightly do inherit heaven's graces
And husband nature's riches from expense;
They are the lords and owners of their faces . . .
—William Shakespeare

IV. God and Mountaintop Removal

by Loyal Jones

The Old Regular Baptists of Eastern Kentucky and elsewhere have made a strong point of holding on to their traditions. Their theology reflects the early Baptists of England and Wales, and each church meets only once a month in the old frontier fashion of congregations sharing scarce preachers. They sing hymns written by such 18th-century hymnists as William Williams, of Wales, and Isaac Watts and John Newton, of England. These hymns are sung in unison, in the lined-out tradition and without instrumental accompaniment. The Primitive and United Baptists maintain similar traditions.

Some Eastern Kentucky churches have been far ahead of other institutions in race relations. Several Primitive and Old Regular Baptist associations had black members through the years. Three congregations were always interracial: the Town Mountain Baptist Church near Hazard, the Little Home Old Regular Baptist Church in Red Fox, and the Church of God Militant Pillar and Ground of Truth, in McRoberts.

So, behind the persistent stereotype about Appalachian people, there is a strong culture. Its values put a priority on religion, family, humility, personal relations, neighborliness, and—certainly—patriotism. Yet "progress" has wrought alterations in the region and its people. Some changes, such as dams and the lakes they create, destroy the communities that the water covers. Likewise, an interstate going through, or a vast strip mine, or an industrial disaster such the Martin County coal-sludge spill, may be so disruptive that families or whole communities do not ever recover.

The present-day practice of blasting whole ranges of mountains and spreading them into the hollows and around the edges of such gigantic coal-recovery operations is bringing profound change to the landscape and to the culture of Eastern Kentucky. There is fear and apprehension caused by powerful explosions and instances of flying rock, gigantic earth-moving equipment and overloaded coal trucks. There is a loss of drinkable water as cracked aquifers are drained away or heavily silted. There is also the loss of a sense of familiarity with one's surroundings that brings disorientation.

With the dwindling supplies of fossil fuels, it is a given that most pools of oil and gas and significant seams of coal will be exploited. But what is troubling in this age of energy scarcity is that the federal government is moving to dismantle the protections for local people and the environment. The government feels no call for conservation or incentives for alternative sources of energy. The result of such policies is damage not only to the culture of Appalachia but also to the values of the entire nation.

> The wolf also shall dwell with the lamb, and the
> leopard shall lie down with the kid; and the calf
> and the young lion and the fatling together; and
> a little child shall lead them. . . .
> They shall not hurt nor destroy in all my holy
> mountain: for the earth shall be full of the
> knowledge of the LORD, as the waters cover the sea.
> —Isaiah 11

Getting Quiet

by Whitney Baker, answering
a column by Kentucky Coal
Association president Bill Caylor

"Mountaintop-removal mining affects a very small percentage of the mountains," Bill Caylor said. "It maximizes coal recovery. It pays the land-owner handsomely. It leaves valuable flat land for future generations. . . ." Caylor also said that the "authors' statement" (printed in Part 1 of this book) "was an emotional tirade playing fast and loose with statements of facts. These are the same people who would be outraged if they knew where their ground beef came from."

Growing up in Middlesboro, I looked out my bedroom window to the west to the bare mountains. They seemed lunar, and very distant. The defacement didn't seem wrong or not wrong. At nights I confused the work lights up there with stars; I noticed that I did that. I was a boy. I wanted everything. My salvation was somewhere in the Sears wish book, not in some blue shape leaning up against the sky.

Things changed for me. I got a car; I started driving into those blue hills. Then I got out of my car and went into the forest on foot. I got a tent and started sleeping there. I got rid of the tent and slept with the moon on my face. It was the right thing to do; I needed the perspective. If you know what the oldest places can do for you, you are grateful for them.

But if you are grown, and still greedy, still hunched over your toy box wishing for something new, you are not grateful. We are not the landowners handsomely paid. We are the families at the bottom of the mountain, the valley where the rusty water runs. We are the pill-eaters, the Sunday-school teachers, the librarians and the Stephen King readers weeding around our patches of corn, or pot, or patch of trillium we swiped from your path. We live here. We love here.

The Bill Caylors of the world believe the best thing to do is to get Appalachia as flat as we can as fast as we can: "valuable flat land for future generations." I encourage everyone to visit, in his or her mind's eye, Caylor's world. Shifty, slatey land covered with hamburger stands. Is Bill going to raise his hand when razing the next mountain, constituting "a large percentage of mountains?" Or did he mean "only affects a very small percentage of remaining mountains?" Hmm, Bill?

We go to the mountains to get married, to go to church, to get alone. No one goes to a slaughterhouse for that. Eastern Kentucky's economic future (not that many of our self-sustaining folk need much economic aid; Lord knows they aren't getting it now), ultimately, is in tourism, or in sustainable industry—not in the next Wal-Mart parking lot. Mr. Caylor says we writers are engaged in an "emotional tirade." No more of a tirade than Mr. Caylor's reaction should we bulldoze his lawn looking for Honey Buns.

We who stand against mountaintop-removal mining are not doing so arbitrarily. We are not deceived about what is necessary for the good of all. There is no emergency to "maximize coal production." The beauty of Appalachia is that it is free and perfect without us. It is our teacher. That is why these mountains are revered around the world. These mountains are quieter than we can be, and larger. Bill, get out of the bulldozer, put down your wish book. Get quiet.

A stewardship, a trust
reposed in us; for which we
must give an account at the
day when our Lord shall call.
—Matthew 25:14

Learning to See

by Gwyn Hyman Rubio

Lately, several experiences have confirmed what I have always felt to be true: the way we learn to see the world is one of the most essential parts of our education. Whitney Baker's article, "Getting Quiet," reminds me of the metamorphosis that we all go through. As a boy growing up in Middlesboro, Baker looked out of his bedroom window upon the bare mountains, which seemed "lunar and very distant," their defacement not affecting him, but as he grew older, his way of seeing the mountains changed. He writes, "I got a car: I started driving into those hills . . . I got a tent and started sleeping there. I got rid of the tent and slept with the moon on my face . . . I needed the perspective."

A few weeks ago, as a member of the Kentucky authors' mountaintop-removal tour, I listened to a man at Hindman Settlement School say much the same thing. As a child, he said, he had cooled off in the creek on his family's mountain property. He had played beneath the tall trees and run through the pasture-land grasses. "You don't see it now, but you are living in paradise," his daddy had told him. Now grown, with his eyes wide open, the man acknowledged that his father had been right. Their mountain was a paradise, he said, but now this paradise was being destroyed.

Just last month, I read Penelope Lively's *Oleander, Jacaranda*—a vivid, poignant account of her unusual childhood growing up English in Egypt during the 1930s and '40s. After a particularly stunning section in which she describes the Egyptian landscape, she addresses the problem of writing a childhood memoir from an adult perspective, admitting that at age ten, when she had traveled on a steamer down the Nile River, she hadn't really seen the unique charm of the land around her. She writes, "A perception of landscape is something learned—it depends upon individual knowledge and experience."

I was born in the flat land of Georgia. As a child, I grew up amidst the longleaf pines of the coastal plains, the loud scent of pine gum burning my nostrils. I walked among rows of striped watermelons in huge, flat fields, picked up nuts in large pecan groves, and went swimming in Lake Blackshear—oblivious to the rich farmland beneath its murky surface, unaware that it had been formed by the damming of the Flint River years ago. When one is a child, a river is simply a river—a place for fun and swimming, not part of something grand that makes a whole. An ancient longleaf is just another pine tree. Unknown to the child is the land's history—the turpentine industry and the subsequent destruction of the old-growth, longleaf pine forests. A pecan grove is merely a place where one can pick up nuts that will go into a pie later, the lovely serenity beneath the trees unnoticed. I was too young to see the natural world through the lens of experience and knowledge.

When I was 26 years old, I moved to Kentucky. It was 1976. My husband and I were recently returned Peace Corps volunteers, and we wanted to do something special with our savings. For this reason, we had traveled throughout the Southeast, looking for the most beautiful land we could afford. We found it in the hills of Wayne County. In these Kentucky hills, my metamorphosis occurred. We built a log cabin on twenty acres of forest land, ten miles from the county seat. A retired coal miner, suffering from black lung, lived down the winding road from us. Every few days he would come to visit. He felt close to the land and was proud to let me know it.

We explored the hills together. He taught me the names of the wildflowers that grew there—trillium, jack-in-the-pulpit, May apple. He pointed out the ginseng and the wild plants that were edible. Under his supervision, I cooked pokeweed for the first time and, much to my amazement, found that it was every bit as delicious as the turnip greens I had eaten as a child. That old miner showed me how to see the land around me, and finally I saw it through wise, grownup eyes.

I married a miner myself. I had ten children. I've got seven now; thirty-one grandchildren and eight great-grand-children. And I'm happy to say not a one's ever crossed a picket line.
—Florence Reese

And so, it was with adult eyes that I looked upon the desecration of our mountains during the authors' tour of Eastern Kentucky and was shocked by what I saw. Miles upon miles of wasteland. Ravaged ridges, bald and barren, or dotted with strange-looking trees. "What kind of tree is that?" I asked the man sitting next to me in the van. "A Russian olive tree," he told me. "Brought in because nothing else will grow." Where were the wildflowers of Kentucky? Where was the pokeweed I had been taught to see? Where were the poplars and the redbuds, native to our mountains? Along these ridges, they were gone. If we, as Kentuckians, close our eyes to the bleeding mountains around us, then mountaintop removal will continue at an alarming pace—and our children will be deprived of their heritage, and of the possibility for their own metamorphosis. If they are robbed of their mountains, their moral, aesthetic, and spiritual education will be stunted, and they will never learn how to see the richness of the natural world around them. Even worse, they will not know what has been taken from them, for parking lots on leveled mountains will become their new landscape, and as victims of this shell game, touted as economic development, they will discover too late what we traded their paradise for.

Visiting the Site of One of the First Churches My Grandfather Pastored

by Tony Crunk

My mother said later that, to the shovel operators, we must have looked like some delegation from out of town that couldn't find the picnic. Or else the funeral. Not so bad my brother and me jumping the fence, and my father, but then my mother, and all of us helping my grandfather over, and finally my grandmother deciding she wanted to see, too.

Then all of us standing together at the rim of the pit in our Sunday clothes, sun reflecting off my grandmother's black patent purse, a few trees still hanging on nearby, roots exposed, like tentacles, like the earth was shrinking under them. The smell of sulphur.

The giant bucket scoops up through the rocks and dirt, the shovel swings around, the bucket empties, and the whole thing swings back, the noise taking an extra second to reach us. I am watching the two men inside, expecting them to notice us, to wave us away because we don't belong there, but they don't. They must be used to it.

Years later, I will remember my grandfather saying that they strip away the land but all they put back is the dirt. Maybe plant a few scrub pine. "Good for nothing anymore," he says now, turning to go back to the car, "except holding the rest of the world together."

It looks almost blue in the sun, the piece of coal I have picked up to take home for a souvenir.

They weren't doing nothing
with that land anyway.
                    —overheard

Big Bertha Stories

fiction by Bobbie Ann Mason

Donald is home again, laughing and singing. He comes home from Central City, near the strip mines, only when he feels like it—like an absentee landlord checking on his property. He is always in such a good humor when he returns that Jeannette forgives him. She cooks for him—ugly, pasty things she gets with food stamps. Sometimes he brings steaks and ice cream, occasionally money. Rodney, their child, hides in the closet when he arrives, and Donald goes around the house talking loudly about the little boy named Rodney who used to live there—the one who fell in the septic tank, or the one stolen by gypsies. The stories change. Rodney usually stays in the closet until he has to pee, and then he hugs his father's knees, forgiving him, just as Jeannette does. The way Donald saunters through the door, swinging a six-pack of beer, with a big grin on his face, takes her breath away. He leans against the door facing, looking sexy in his baseball cap and his shaggy red beard and his sun-glasses. He wears sunglasses to be like the Blues Brothers. I should have my head examined, Jeannette thinks.

The last time Donald was home, they went to the shopping center to buy Rodney some shoes advertised for sale. They stayed at the shopping center half the afternoon, just looking around. Donald and Rodney played video games. Jeannette felt they were a normal family. Then, in the parking lot, they stopped to watch a man on a platform demonstrating snakes. Children were petting a twelve-foot python coiled around the man's shoulders. Jeannette felt faint.

"Snakes won't hurt you unless you hurt them," said Donald as Rodney stroked the snake.

"It feels like chocolate," Rodney said.

The snake man took a tarantula from a plastic box and held it lovingly in his palm. He said, "If you drop a tarantula, it will shatter like a Christmas ornament."

"I hate this," said Jeannette.

"Let's get out of here," said Donald.

Jeannette felt her family disintegrating like a spider shattering as Donald hurried them away from the shopping center. Rodney squalled and Donald dragged him along. Jeannette wanted to stop for ice cream. She wanted them all to sit quietly together in a booth, but Donald rushed them to the car, and he drove them home in silence, his face growing grim.

"Did you have bad dreams about the snakes?" Jeannette asked Rodney the next morning at breakfast. They were eating pancakes made with generic pancake mix. Rodney slapped his fork in the pond of syrup on his pancakes. "The black racer is the farmer's friend," he said soberly, repeating a fact learned from the snake man.

"Big Bertha kept black racers," said Donald. "She trained them for the 500." Donald doesn't tell Rodney ordinary children's stories. He tells him a series of strange stories he makes up about Big Bertha. Big Bertha is what he calls the huge strip-mining machine in Muhlenberg County, but he has Rodney believing that Big Bertha is a female version of Paul Bunyan.

"Snakes don't run in the 500," said Rodney.

"This wasn't the Indy 500," said Donald. "This was the Possum Trot 500, and it was a long time ago. Big Bertha started the original 500, with snakes. Black racers and blue racers mainly. Also some red-and-white-striped racers, but those are rare."

"We always ran for a hoe if we saw a black racer," Jeannette said, remembering her childhood in the country.

In a way, Donald's absences are a fine arrangement, even considerate. He is sparing them his darkest moods, when he can't cope with his memories of Vietnam. Vietnam had never seemed such a meaningful fact until a couple of years ago, when he grew depressed and moody, and then he started going away to Central City. He frightened Jeannette, and she always said the wrong thing in her efforts to soothe him. If the welfare people find out he is spending occasional weekends at home, and even bringing some money, they will cut off her assistance. She applied for welfare because she can't depend on him to send money, but she knows he blames her for losing faith in him.

He isn't really working regularly at the strip mines. He is mostly just hanging around there, watching the land being scraped away, trees coming down, bushes flung in the air. Sometimes he operates a steam shovel, and when he comes home his clothes are filled with clay and it is caked on his shoes. The clay is the color of butterscotch pudding.

At first, he tried to explain himself to Jeannette. He said, "If we could have had tanks over there as big as Big Bertha, we wouldn't have lost the war. Strip mining is just like what we were doing over there. We were stripping off the top. The topsoil is like the culture and the people, the best part of the land and the country. America was just stripping off the top, the best. We ruined it. Here, at least the coal companies have to plant vetch and loblolly pines and all kinds of trees and bushes. If we'd done that in Vietnam, maybe we'd have left that country in better shape."

"Wasn't Vietnam a long time ago?" Jeannette asked.

She didn't want to hear about Vietnam. She thought it was unhealthy to dwell on it so much. He should live in the present. Her mother is afraid Donald will do something violent, because she once read in a newspaper that a vet in Louisville held his little girl hostage in their apartment until he had a shootout with the police and was killed. But Jeannette can't imagine Donald doing anything so extreme. When she first met him, several years ago, at her parents' pit-barbecue luncheonette where she was working then, he had a good job at a lumberyard and he dressed nicely. He took her out to eat at a fancy restaurant. They got plastered and ended up at a motel in Tupelo, Mississippi, on Elvis Presley Boulevard. Back then, he talked nostalgically about his year in Vietnam, about how beautiful it was, how different the people were. He could never seem to explain what he meant. "They're just different," he said.

They went riding around in a yellow 1957 Chevy convertible. He drives too fast now, but he didn't then, maybe because he was so protective of the car. It was a classic. He sold it three years ago and made a good profit. About the time he sold the Chevy, his moods began changing, his even-tempered nature shifting, like driving on a smooth interstate and then switching to a secondary road. He had headaches and bad dreams. But his nightmares seemed trivial. He dreamed of riding a train through the Rocky Mountains, of hijacking a plane to Cuba, of stringing up barbed wire around the house. He dreamed he lost a doll. He got drunk and rammed the car, the Chevy's successor, into a Civil War statue in front of the courthouse. When he got depressed over the meaninglessness of his job, Jeannette felt guilty about spending money on something nice for the house, and she tried to make him feel his job had meaning by reminding him that, after all, they had a child to think of. "I don't like his name," Donald said once. "What a stupid name. Rodney. I never did like it."

Rodney has dreams about Big Bertha, echoes of his father's nightmares, like TV cartoon versions of Donald's memories of the war. But Rodney loves the stories, even though they are confusing, with lots of loose ends. The latest in the Big Bertha series is "Big Bertha and the Neutron Bomb." Last week it was "Big Bertha and the MX Missile." In the new story, Big Bertha takes a trip to California to go surfing with Big Mo, her male counterpart. On the beach, corn dogs and snow cones are free and the surfboards turn into dolphins. Everyone is having fun until the neutron bomb comes. Rodney loves the part where everyone keels over dead. Donald acts it out, keeling over on the rug. All the dolphins and the surfers keel over, everyone except Big Bertha. Big Bertha is so big she is immune to the neutron bomb.

"Those stories aren't true," Jeannette tells Rodney.

Rodney staggers and falls down on the rug, his arms and legs akimbo. He gets the giggles and can't stop. When his spasms finally subside, he says, "I told Scottie Bidwell about Big Bertha and he didn't believe me."

Donald picks Rodney up under the armpits and sets him upright. "You tell Scottie Bidwell if he saw Big Bertha he would pee in his pants on the spot, he would be so impressed."

"Are you scared of Big Bertha?"

"No, I'm not. Big Bertha is just like a wonderful woman, a big fat woman who can sing the blues. Have you ever heard Big Mama Thornton?"

"No."

"Well, Big Bertha's like her, only she's the size of a tall building. She's slow as a turtle and when she crosses the road they have to reroute traffic. She's big enough to straddle a four-lane highway. She's so tall she can see all the way to Tennessee, and when she belches, there's a tornado. She's really something. She can even fly."

"She's too big to fly," Rodney says doubtfully. He makes a face like a wadded-up washrag and Donald wrestles him to the floor again.

Donald has been drinking all evening, but he isn't drunk. The ice cubes melt and he pours the drink out and refills it. He keeps on talking. Jeannette cannot remember him talking so much about the war. He is telling her about an ammunition dump. Jeannette has the vague idea that an ammo dump is a mound of shotgun shells, heaps of cartridge casings and bomb shells, or whatever is left over, a vast waste pile from the war, but Donald says that is wrong. He has spent an hour describing it in detail, so that she will understand.

He refills the glass with ice, some 7-Up, and a shot of Jim Beam. He slams doors and drawers, looking for a compass. Jeannette can't keep track of the conversation. It doesn't matter that her hair is uncombed and her lipstick eaten away. He isn't seeing her.

"I want to draw the compound for you," he says, sitting down at the table with a sheet of Rodney's tablet paper.

Donald draws the map in red and blue ballpoint, with asterisks and technical labels that mean nothing to her. He draws some circles with the compass and measures some angles. He makes a red dot on an oblique line, a path that leads to the ammo dump.

"That's where I was. Right there," he says. "There was a water buffalo that tripped a land mine and its horn just flew off and stuck in the wall of the barracks like a machete thrown backhanded." He puts a dot where the land mine was, and he doodles awhile with the red ballpoint pen, scribbling something on the edge of the map that looks like feathers. "The dump was here and I was there and over there was where we piled the sandbags. And here were the tanks." He draws tanks, a row of squares with handles—guns sticking out.

"Why are you going to so much trouble to tell me about a buffalo horn that got stuck in a wall?" she wants to know.

But Donald just looks at her as though she has asked something obvious.

"Maybe I could understand if you'd let me," she says cautiously.

"You could never understand." He draws another tank. In bed, it is the same as it has been since he started going away to Central City—the way he claims his side of the bed, turning away from her. Tonight, she reaches for him and he lets her be close to him. She cries for a while and he lies there, waiting for her to finish, as though she were merely putting on makeup.

"Do you want me to tell you a Big Bertha story?" he asks playfully.

"You act like you're in love with Big Bertha."

He laughs, breathing on her. But he won't come closer.

"You don't care what I look like anymore. What am I supposed to think?"

"There's nobody else. There's not anybody but you."

Loving a giant machine is incomprehensible to Jeannette. There must be another woman, someone that large in his mind. Jeannette has seen the strip-mining machine. The top of the crane is visible beyond a rise along the parkway. The strip mining is kept just out of sight of travelers because it would give them a poor image of Kentucky.

For three weeks, Jeannette has been seeing a psychologist at the free mental health clinic. He's a small man from out of state. His name is Dr. Robinson, but she calls him The Rapist, because the word *therapist* can be divided into two words, the rapist. He doesn't think her joke is clever, and he acts as if he has heard it a thousand times before. He has a habit of saying "Go with that feeling," the same way Bob Newhart did on his old TV show. It's probably the first lesson in the textbook, Jeannette thinks.

She told him about Donald's last days on the job at the lumberyard—how he let the stack of lumber fall deliberately and still didn't know why, and how he went away soon after that, and how the Big Bertha stories started. Dr. Robinson seems to be waiting for her to make something out of it all, but it's maddening that he won't tell her what to do. After three visits, Jeannette has grown angry with him, and now she's holding back things. She won't tell him whether Donald slept with her or not when he came home last. Let him guess, she thinks.

"Talk about yourself," he says.

"What about me?"

"You speak so vaguely about Donald that I get the feeling you see him as somebody larger than life. I can't quite picture him. That makes me wonder what that says about you." He touches the end of his tie to his nose and sniffs it.

When Jeannette suggests that she bring Donald in, the therapist looks bored and says nothing. "He had another nightmare when he was home last," Jeannette says. "He dreamed he was crawling though tall grass and people were after him." "How do you feel about that?" the Rapist asks eagerly. "I didn't have the nightmare," she says coldly. "Donald did. I came to you to get advice about Donald, and you're acting like I'm the one who's crazy. I'm not crazy. But I'm lonely."

Jeannette's mother, behind the counter of the luncheonette, looks lovingly at Rodney pushing buttons on the jukebox in the corner. "It's a shame about that youngun," she says tearfully. "That boy needs a daddy."

"What are you trying to tell me? That I should file for a divorce and get Rodney a new daddy?"

Her mother looks hurt. "No, honey," she says. "You need to get Donald to seek the Lord. And you need to pray more. You haven't been to church lately."

"Have some more barbecue," Jeannette's father booms, as he comes in from the back kitchen. "And I want you to take a pound home with you. You've got a growing boy to feed."

"I want to take Rodney to church," Mama says. "I want to show him off, and it might do him some good."

"People will think he's an orphan," Dad says.

"I don't care," Mama says. "I just love him to pieces and I want to take him to church. Do you care if I take him to church, Jeannette?"

"No, I don't care if you take him to church." She takes the pound of barbecue from her father. Grease splotches the brown wrapping paper. Dad has given them so much barbecue that Rodney is burned out on it and won't eat it anymore.

Jeannette wonders if she would file for divorce if she could get a job. It is a thought—for the child's sake, she thinks. But there aren't many jobs around. With the cost of a babysitter, it doesn't pay her to work. When Donald first went away, her mother kept Rodney and she had a good job, waitressing at a steak house, but the steak house burned down one night—a grease fire in the kitchen. After that, she couldn't find a steady job, and she was reluctant to ask her mother to keep Rodney again because of her bad hip. At the steak house, men gave her tips and left their telephone numbers on the bill when they paid. One note said, "I want to hold your muffins." They were real-estate developers and businessmen on important missions for the Tennessee Valley Authority. They were boisterous and they drank too much. They said they'd take her on a cruise on the Delta Queen, but she didn't believe them. She knew how expensive it was. They talked about their speed boats and invited her for rides on Lake Barkley, or for spins in their private planes. They always used the word spin. The idea made her dizzy. Once, Jeannette let an electronics salesman take her for a ride in his Cadillac, and they breezed down the wilderness road through the Land Between the Lakes. His car had automatic windows and a stereo system and lighted computer-screen numbers on the dash that told him how many miles to the gallon he was getting and other statistics. He said the numbers distracted him and he had almost had several wrecks. At the restaurant, he had been flamboyant, admired by his companions. Alone with Jeannette in the Cadillac, on the Trace, he was shy and awkward, and really not very interesting. The most interesting thing about him, Jeannette thought, was all the lighted numbers on his dash-board. The Cadillac had everything but video games. But she'd rather be riding around with Donald, no matter where they ended up.

While the social worker is there, filling out her report, Jeannette listens for Donald's car. When the social worker drove up, the flutter and wheeze of her car sounded like Donald's old Chevy, and for a moment Jeannette's mind lapsed back in time. Now she listens, hoping he won't drive up. The social worker is younger than Jeannette and has been to college. Her name is Miss Bailey, and she is excessively cheerful, as though in her line of work she has seen hardships that make Jeannette's troubles seem like a trip to Hawaii.

"Is your little boy still having those bad dreams?" Miss Bailey asks, looking up from her clipboard.

Jeannette nods and looks at Rodney, who has his finger in his mouth and won't speak.

"Has the cat got your tongue?" Miss Bailey asks.

"Show her your pictures, Rodney." Jeannette explains, "He won't talk about the dreams, but he draws pictures of them."

Rodney brings his tablet of pictures and flips through them silently. Miss Bailey says "Hmm." They are stark line drawings, remarkably steady lines for his age. "What is this one?" she asks. "Let me guess. Two scoops of ice cream?"

The picture is two huge circles, filling the page, with three tiny stick figures in the corner.

"These are Big Bertha's titties," says Rodney.

Miss Bailey chuckles and winks at Jeannette. "What do you like to read, hon?" she asks Rodney.

"Nothing."

"He can read," Jeannette says. "He's smart."

"Do you like to read?" Miss Bailey asks Jeannette. She glances at the pile of paperbacks on the coffee table. She is probably going to ask Jeannette where she got the money for them.

"I don't read," Jeannette says. "If I read, I just go crazy."

When she told The Rapist she couldn't concentrate on anything serious, he said she read romance novels in order to escape from reality. "Reality, hell!" she had said. "Reality's my whole problem."

"It's too bad Rodney's not here," Donald is saying. Rodney is in the closet again. "Santa Claus has to take back all these toys. Rodney would love this bicycle! And this Pac-Man game. Santa has to take back so many things he'll have to have a pickup truck!"

"You didn't bring him anything. You never bring him anything," says Jeanette.

He has brought doughnuts and dirty laundry. The clothes he is wearing are caked with clay. His beard is lighter, from working out in the sun, and he looks like his usual joyful self, the way he always is before his moods take over, like migraine headaches, which some people describe as storms.

Donald coaxes Rodney out of the closet with the doughnuts.

"Were you a good boy this week?"

"I don't know."

"I hear you went to the shopping center and showed out." It is not true that Rodney made a big scene. Jeannette has already explained that Rodney was upset because she wouldn't buy him an Atari. But she didn't blame him for crying. She was tired of being unable to buy him anything.

Rodney eats two doughnuts and Donald tells him a long, confusing story about Big Bertha and a rock-and-roll band. Rodney interrupts him with dozens of questions. In the story, the rock-and-roll band gives a concert in a place that turns out to be a toxic-waste dump and the contamination is spread all over the country. Big Bertha's solution to this problem is not at all clear. Jeannette stays in the kitchen, trying to think of something original to do with instant potatoes and leftover barbecue.

"We can't go on like this," she says that evening in bed.

"We're just hurting each other. Something has to change."

He grins like a kid. "Coming home from Muhlenberg County is like R and R—rest and recreation. I explain that in case you think R and R means rock and roll. Or maybe rumps and rears. Or rust and rot." He laughs and draws a circle in the air with his cigarette.

"I'm not that dumb."

"When I leave, I go back to the mines." He sighs, as though the mines were some eternal burden.

Her mind skips ahead to the future: Donald locked away somewhere, coloring in a coloring book and making clay pots, her and Rodney in some other town, with another man—someone dull and not at all sexy. Summoning up her courage, she says, "I haven't been through what you've been through and maybe I don't have a right to say this, but sometimes I think you act superior because you went to Vietnam, like nobody can ever know what you know. Well, maybe not. But you've still got your legs, even if you don't know what to do with what's between them anymore." Bursting into tears of apology, she can't help adding, "You can't go on telling Rodney those awful stories. He has nightmares when you're gone."

Donald rises from bed and grabs Rodney's picture from the dresser, holding it as he might a hand grenade.

"Kids betray you," he says, turning the picture in his hand.

"If you cared about him, you'd stay here." As he sets the picture down, she asks, "What can I do? How can I understand what's going on in your mind? Why do you go there? Strip mining's bad for the ecology and you don't have any business strip mining."

"My job is serious, Jeannette. I run that steam shovel and put the topsoil back on. I'm reclaiming the land." He keeps talking, in a gentler voice, about strip mining, the same old things she has heard before, comparing Big Bertha to a supertank. If only they had had Big Bertha in Vietnam. He says, "When they strip the top off, I keep looking for those tunnels where the Viet Cong hid. They had so many tunnels it was unbelievable. Imagine Mammoth Cave going all the way across Kentucky."

"Mammoth Cave's one of the natural wonders of the world," says Jeannette brightly. She is saying the wrong thing again.

At the kitchen table at 2 a.m., he's telling about C-5A's. A C-5A is so big it can carry troops and tanks and helicopters, but it's not big enough to hold Big Bertha. He rambles on, and when Jeannette shows him Rodney's drawings of the circles, Donald smiles. Dreamily, he begins talking about women's breasts and thighs—the large, round thighs and big round breasts of American women, contrasted with the frail, delicate beauty of the Orientals. It is like comparing oven broilers and banties, he says. Jeannette relaxes. A confession about another lover from long ago is not so hard to take. He seems stuck on the breasts and thighs of American women—insisting that she understand how small and delicate the Orientals are, but then he abruptly returns to tanks and helicopters.

"A Bell Huey Cobra—my God, what a beautiful machine. So efficient!" Donald takes the food processor blade from the drawer where Jeannette keeps it. He says, "A rotor blade from a chopper could just slice anything to bits."

"Don't do that," Jeannette says.

He is trying to spin the blade on the counter, like a top. "Here's what would happen when a chopper blade hits a power line—not many of those over there!—or a tree. Not many trees, either, come to think of it, after all the Agent Orange." He drops the blade and it glances off the open drawer and falls to the floor, spiking the vinyl.

At first, Jeannette thinks the screams are hers, but they are his. She watches him cry. She has never seen anyone cry so hard, like an intense summer thundershower. All she knows to do is shove Kleenex at him. Finally, he is able to say, "You thought I was going to hurt you. That's why I'm crying."

"Go ahead and cry," Jeannette says, holding him close.

"Don't go away."

"I'm right here. I'm not going anywhere."

In the night, she still listens, knowing his monologue is being burned like a tattoo onto her brain. She will never forget it. His voice grows soft and he plays with a ballpoint pen, jabbing holes in a paper towel. Bullet holes, she thinks. His beard is like a bird's nest, woven with dark corn silks.

"This is just a story," he says. "Don't mean nothing. Just relax." She is sitting on the hard edge of the kitchen chair, her toes cold on the floor, waiting. His tears have dried up and left a slight catch in his voice.

"We were in a big camp near a village. It was pretty routine and kind of soft there for a while. Now and then we'd go into Da Nang and whoop it up.

We had been in the jungle for several months, so the two months at this village was a sort of rest—an R and R almost. Don't shiver. This is just a little story. Don't mean nothing! This is nothing, compared to what I could tell you. Just listen. We lost our fear. At night there would be some incoming and we'd see these tracers in the sky, like shooting stars up close, but it was all pretty minor and we didn't take it seriously, after what we'd been through. In the village I knew this Vietnamese family—a woman and her two daughters. They sold Cokes and beer to the GIs. The oldest daughter was named Phan. She could speak a little English. She was really smart. I used to go see them in their hooch in the afternoons—in the siesta time of the day. It was so hot there.

"Phan was beautiful, like the country. The village was ratty, but the country was pretty. And she was beautiful, like she had grown up out of the jungle, like one of those flowers that bloomed high up in the trees and freaked us out some-times, thinking it was a sniper. She was so gentle, with those eyes shaped like peach pits, and she was no bigger than a child of maybe thirteen or fourteen. I felt funny about her size at first, but later it didn't matter. It was just some wonderful feature about her, like a woman's hair, or her breasts."

He stops and listens, the way they used to listen for crying sounds when Rodney was a baby. He says, "She'd take those big banana leaves and fan me while I lay there in the heat."

"I didn't know they had bananas over there."

"There's a lot you don't know! Listen! Phan was 23, and her brothers were off fighting. I never even asked which side they were fighting on." He laughs. "She got a kick out of the word fan. I told her than fan was the same word as her name. She thought I meant her name was banana. In Vietnamese, the same word can have a dozen different meanings, depending on your tone of voice. I bet you didn't know that, did you?"

"No. What happened to her?"

"I don't know."

"Is that the end of the story?"

"I don't know." Donald pauses, then goes on talking about the village, the girl, the banana leaves, talking in a monotone that is making Jeannette's flesh crawl. He could be the news radio from the next room.

"You must have really liked that place. Do you wish you could go back there and find out what happened to her?"

"It's not there anymore," he says. "It blew up."

Donald abruptly goes to the bathroom. She hears the water running, the pipes in the basement shaking.

"It was so pretty," he says when he returns. "That jungle was the most beautiful place in the world. You'd have thought you were in paradise. But we blew it sky-high."

In her arms, he is shaking, like the pipes in the basement, which are still vibrating. Then the pipes let go, after a long shudder, but he continues to tremble.

They are driving to the Veterans Hospital. It was Donald's idea. She didn't have to persuade him. When she made up the bed in the morning—with a finality that shocked her, as though she knew they wouldn't be in it again together—he told her he would like R and R. Rest was what he needed.

Neither of them had slept at all during the night. Jeannette felt she had to stay awake, to listen for more.

"Talk about strip mining," she says now. "That's what they'll do to your head. They'll dig out all those ugly memories, I hope. We don't need them around here anymore." She pats his knee.

It is a cloudless day, not the setting for a sober journey. She drives and Donald goes along obediently, with the resignation of an old man being taken to a rest home. They are driving through southern Illinois, known as Little Egypt, for some obscure reason Jeannette has never understood. Donald still talks, but very quietly, without urgency. When he mentions scenery, Jeannette thinks of the early days of their marriage, when they would take a drive like this and laugh hysterically. Now Jeannette points out funny things they see. The Little Egypt Hot Dog World, Pharaohs Cleaners, Pyramid Body Shop. She is scarcely aware that she is driving, and when she sees a sign, LITTLE EGYPT STARLITE CLUB, she is confused for a moment, wondering where she has been transported.

As they park, he asks, "What will you tell Rodney if I don't come back? What if they keep me here indefinitely?"

"You're coming back. I'm telling him you're coming back soon."

"Tell him I went off with Big Bertha. Tell him she's taking me on a sea cruise, to the South Seas."

He starts singing "Sea Cruise." He grins and pokes her in the ribs.

"You're coming back," she says.

Donald writes from the VA Hospital, saying that he is making progress. They are running tests, and he meets in a therapy group in which all the veterans trade memories. Jeannette is no longer on welfare because now she has a job waitressing at Fred's Family Restaurant. She waits on families, waits for Donald to come home so they can come here and eat together like a family. The fathers look at her with downcast eyes, and the children throw food. While Donald is gone, she rearranges the furniture. She reads some books from the library. She does a lot of thinking. It occurs to her that even though she loved him, she has thought of Donald primarily as a husband, a provider, someone whose name she shared, the father of her child, someone like the fathers who came to the Wednesday-night-all-you-can-eat fish fry. She hasn't thought of him as himself. She wasn't brought up that way, to examine someone's soul. When it comes to something deep inside, nobody will take it out and examine it, the way they will look at clothing in a store for flaws in the manufacturing. She tries to explain all this to The Rapist, and he says she's looking better, got sparkle in her eyes. "Big deal," says Jeannette. "Is that all you can say?"

She takes Rodney to the shopping center, their favorite thing to do together, even though Rodney always begs to buy something. They go to Penney's perfume counter. There, she usually hits a sample bottle of cologne—Chantilly or Charlie or something strong. Today she hits two or three and comes out of Penney's smelling like a flower garden.

"You stink!" Rodney cries, wrinkling his nose like a rabbit.

"Big Bertha smells like this, only a thousand times worse, she's so big," says Jeannette impulsively. "Didn't Daddy tell you that?"

"Daddy's a messenger from the devil."

This is an idea he must have gotten from church. Her parents have been taking him every Sunday. When Jeannette tries to reassure him about his father, Rodney is skeptical. "He gets that funny look on his face like he can see through me," the child says.

"Something's missing," Jeannette says, with a rush of optimism, a feeling of recognition. "Something happened to him once and took out the part that shows how much he cares for us."

"The way we had the cat fixed?"

"I guess. Something like that." The appropriateness of his remark stuns her, as though, in a way, her child has understood Donald all along. Rodney's pictures have been more peaceful lately, pictures of skinny trees and airplanes flying low. This morning, he drew pictures of tall grass, with creatures hiding in it. The grass is tilted at an angle, as though a light breeze is blowing through it.

With her paycheck, Jeannette buys Rodney a present, a miniature trampoline they have seen advertised on television. It is called Mr. Bouncer. Rodney is thrilled about the trampoline, and he jumps on it until his face is red. Jeannette discovers that she enjoys it too. She puts it out on the grass, and they take turns jumping. She has an image of herself on the trampoline, her sailor collar flapping, at the moment when Donald returns and sees her flying.

One day a neighbor driving by slows down and calls out to Jeannette as she is bouncing on the trampoline, "You'll tear your insides loose!" Jeannette starts thinking about that and the idea is so horrifying she stops jumping so much. That night, she has a nightmare about the trampoline. In her dream, she is jumping on soft moss, and then it turns into a springy pile of dead bodies.

Why does a virtuous man take delight in landscapes? Because the din of the dusty world and the locked-in-ness of human habitations are what human nature habitually abhors; while on the contrary, haze, mist, and the haunting spirits of the mountains are what human nature seeks, and yet can rarely find.
　　　　　—Kuo His

Contempt for Small Places

by Wendell Berry

Newspaper editorials deplore such human-caused degradations of the oceans as the Gulf of Mexico's "dead zone," and reporters describe practices like "mountaintop-removal" mining in Eastern Kentucky. Someday we may finally understand the connections.

The health of the oceans depends on the health of rivers; the health of rivers depends on the health of small streams; the health of small streams depends on the health of their watersheds. The health of the water is exactly the same as the health of the land; the health of small places is exactly the same as the health of large places. As we know, disease is hard to confine. Because natural law is in force everywhere, infections move.

We cannot immunize the continents and the oceans against our contempt for small places and small streams. Small destructions add up, and finally they are understood collectively as large destructions. Excessive nutrient runoff from farms and animal factories in the Mississippi watershed has caused, in the Gulf of Mexico, a hypoxic or "dead" zone of five or six thousand square miles. In forty-odd years, strip mining in the Appalachian coalfields, culminating in mountain removal, has gone far toward the destruction of a whole region, with untold damage to the region's people, to watersheds, and to the waters downstream.

There is not a more exemplary history of our contempt for small places than that of Eastern Kentucky coal mining, which has enriched many absentee corporate shareholders and left the region impoverished and defaced. Coal industry representatives are now defending mountaintop removal—and its attendant damage to forests, streams, wells, dwellings, roads, and community life—by saying that in "ten, fifteen, twenty years" the land will be restored, and that such mining has "created the [level] land" needed for further industrial development.

But when you remove a mountain, you also remove the topsoil and the forest, and you do immeasurable violence to the ecosystem and the watershed. These things are not to be restored in ten or twenty years, or in 1,000 or 2,000 years. As for the manufacture of level places for industrial

development, the supply has already far exceeded any foreseeable demand. And the devastation continues.

The contradictions in the state's effort "to balance the competing interests" were stated as follows by Ewell Balltrip, director of the Kentucky Appalachian Commission: "If you don't have mining, you don't have an economy, and if you don't have an economy, you don't have a way for the people to live. But if you don't have environmental quality, you won't create the kind of place where people want to live."

Yes. And if the clearly foreseeable result is a region of flat industrial sites where nobody wants to live, we need a better economy.

Personal Statement

by Daymon Morgan

Horizon Resources, the coal company that works near my home, has trespassed along one of my property lines. They have done damage to the land and to my personal property—trees, rocks, and dirt debris have been pushed onto my property and down the side of the mountain. This damage causes erosion and may even damage the creek at the bottom of the hollow.

The coal industry is an outlaw industry that does not consider the rights of its neighbors or the rights of the land and environment. The industry is out to make a profit and has no regard for the damage done to the citizens of this country.

This trespassing issue is just another example of the coal industry's blatant disregard. Within the last five years, two homes on my property have been damaged from the blasting. I believe that almost everyone up Bad Creek has sued the coal industry at least once for damages done to their property.

And my community is not the only community affected by this outlaw industry. Folks over in Raccoon Creek, in Greasy Creek, Viper, and Vicco also are complaining.

This is not just a private-property issue. It is everyone's problem, and something must be done. The coal industry and our state regulators need to have more respect for the mountains and people of this region.

The coal industry is leaving us destroyed, with no water, no trees, no wildlife habitats, or any economic prospects for our future.

This must stop.

An Ontological Postcard From Cumberland Falls

by Richard Taylor

One thing translates into another.
Rotting shingles pried off the roof
Raise mountains at the landfill.
Windfalls from the front yard
(these rents in the maples)
Plug the gullies of our eroded slopes—
Those contours, this valley, in turn
Inscribed by migrating streams,
The ample ghosts of fickle water
That make orphans of the hills.
Chafed by these currents,
However muddied, however deep,
We melt, one particle at a time.
Now we are water, now silt.

Solomon's Trees

by Steven R. Cope

It killed me when they cut down the trees. It did. I can't go beyond saying that. I spent the next ten years of my life just trying to get over it, to get well, to get better, but here I am in the tenth year and I'm not over it yet. It killed me. I don't know how else to put it.

It's like you spend all of your every day right out under those trees, or looking up at those trees, or just knowing the trees are there. Or it's like you go to school in town and then you go to work in town and you can't wait to get back home so you can walk out under those trees. It's like the trees were you. It's like you were the trees. It's like while they were there you were there and when they were gone you were gone. It's like you wake up and you're gone, everything you were or thought you were. And there's nothing to go out to or to walk under.

"What's wrong, Solomon?"

And you're Solomon. But sometimes you forget that you are.

"Solomon?"

And then you remember. But even when you remember, you don't know now what to say.

"What's wrong with you, Solomon?"

But they wouldn't understand. They'd think the hills and the rivers and the green pastures and the trees could be cut off and blown up and poisoned and paved and life goes on as it was. And they'd think you go on as you were and they go on as they were and nothing is changed, nothing lost, nothing marred or diminished. So you can't tell them about the trees. You just say—

"There's nothing wrong with me."

"Well, you sure act like there is."

"Can I help it how I act?"

And that's as close as you can get to telling them what's wrong—telling them you can't help it how you act. Cause you can't and that's the truth. And you know that you can't. Not now. Used to, you could help it. You could walk out under the trees—the low trees or the high trees, the green trees or the brown trees—and all at once you could help it. The trees would get you to where you could help it. They would help you find a way. Now you can't help it at all. You can't help how you act at all.

"We're gonna play us some ball, Solomon. You with us?"

"I'm not with you," you say. Cause when you're chasing down the ball, you'll still be thinking about the trees. Your feet will be moving across the dry, hard clay, but you won't even know it or care. You'll just be doing it as in a dream, a daze, doing it the way you once might have done.

You'll just pick up the ball and throw it back to the infield and the game will go on quite without you. And when the game is over and somebody

wins, you'll go home and just sit there, sit there without trees, probably still in a daze. And life will go on quite without you.

"What's the matter, Solomon?"

The trees are not big and not tall. The leaves are not little and not green and the wind does not blow and pitch and whirl right out along the branches. And underneath the trees in rain or snow, you are no longer so alive you have to tremble, just stand there and tremble.

"Solomon?"

And you are not attached now to all that is, all that's attached to the trees, all that's linked and intertwined, all that's grown up and meshed together. From the roots, from the leaves, from the gnarled or smooth branches. From the camouflaged earth. The flitting sky. The tree-sounds out of everywhere always lifting your eyes. Or the wild blinking silences. The nests. The dens. And life stirring beneath your feet, your hands, above your head, beneath your own skin . . .

"Solomon!"

"What?"

"We're gonna go play us some baseball. You comin?"

And then you simply must speak. "You all remember the trees?"

And some of them say "Yes."

And you say, "How old were you then, back when there were trees?"

And one of them says, "Just a kid. How old were you?"

"Me?"

So you sit down on the ground and try to figure. You try to figure it in your head, but there's no way now that you can do it right. It's hard to remember when you were there cause when you close your eyes you are there now. The wind in the branches sings and moans and you are sitting by the falls, hills of white oaks high above you, mountains of hemlock, pine, bursts of redbud. Not a saw screams out of anywhere. Not a dragline or a drill. Not a sluice. Not a truck. Not a tree crashes to the earth. Not a baby bird smashes. The frog floats up to the surface, yawns. A woodpecker drums. A red squirrel eases down a limb. And out there in the outfield the poplars rise straight into heaven, up through the hickory and the dogwood, the bees and a lone butterfly . . .

"Solomon?"

And somebody else says, "Where'd he go?"

And someone else, "What's wrong, Solomon?"

And you sit there in the trees and wonder who on earth they could be talking to. Or why they would talk when they didn't have to.

Notes From a Not-Quite-Native Daughter

by Leatha Kendrick

> How can I stand on the ground every day and not feel its power? How can I live my life stepping on this stuff and not wonder at it?
> —*Dirt: The Ecstatic Skin of the Earth* by William Bryant Logan

My concern lies in the top six inches of the earth. Not that I don't love what we once thought were the "everlasting" hills, but my daddy's daddy was a dirt farmer, I've been proud to say all my life, meaning that my paternal grandfather took his living from the soil and supported his ten children and the assorted relatives who lived with them off and on with what he could produce on a succession of rocky western Kentucky farms.

My perspective on mountaintop removal, then, is somewhat different from that of a native mountaineer. I join this fray in awe of the power of topsoil, where between one and ten million microorganisms can live in a gram, invisibly going about the daily business of cleaning up their environment and creating the conditions in which plants can thrive. I come from farming people whose lives and livelihood depended on care of the soil.

More than thirty years ago, when I married Will and moved into the mountains, a brought-on bride from the (relative) flatlands of the Pennyrile part of Kentucky, I had a hard time overcoming my alarm at what seemed to me to be the widespread disregard for soil in Eastern Kentucky. Despite many gardens (particularly in the more remote parts of the Floyd county and along old route 80 in Knott County), most big bottomlands were routinely filled with stone and subsoil to provide space for shopping centers and house seats. What Will saw as a way to raise a house out of the floodplain, I saw as the permanent destruction of scarce alluvial soil. Though he is as much a gardener and lover of plants as I am, we were clearly coming from different perspectives when it came to bottomland. What we did agree on was that the beauty and wholeness of his native place was marred by misuse —not just the mining that scarred the landscape and left its coat of grime on streets and buildings as well as streams, but also a daily disregard for the hills and creeks evidenced in the garbage strewn down hillsides and washed down the watersheds into Dewey Lake, which every spring was choked with plastic milk jugs, disposable diapers, and the detritus of modern life.

Nevertheless, I set in at once to get to know this new landscape, hiking up the road to the cemetery above our rented house on Cow Creek, learning the names of the plants that lined the rutted dirt track. We moved to Eastern Kentucky in August, so the jewelweed was in full bloom when I took those walks, blossoms hanging like tiny orange orchids suspended from succulent, fluid-filled stems. I learned the juice of these stems would prevent and cure poison ivy. I learned that poison ivy often grows near jewelweed—as well as up trees in huge roping, root-covered vines that climbed poplars like snakes. I found small, hidden waterfalls, and unexpected glades of ferns in what I came to think of as the sweet clefts between the hills. I was astonished at the variety of plants and trees that covered the hillsides and sat at eye-level outside my windows.

Later I learned I was living on the edge of the only mixed mesophytic forest in North America. (Even later I discovered that E. Lucy Braun, working out of the University of Cincinnati, had coined this name and that she was the first scientist to recognize that this forest covering the Appalachians was a coherent system—the world's oldest and biologically richest temperate-zone hardwood system.) Ecologists call it the "mother forest." All I knew was that I was finding plants I had never seen before and more kinds of trees than I thought could exist in one place. That is true: More kinds of trees *do* live together in the Appalachian forest than anywhere else in the world. What's more, this old forest literally seeded and repopulated the forests around it after the last glaciers receded, and it still serves as a sort of incubator for many species of plants, as well as a sheltering stopover place for migrating songbirds. That's the other richness Will and I soon cultivated: backyard birds—a huge variety of woodpeckers, finches, titmice, juncos, grosbeaks, in addition to the more familiar cardinals, chickadees, and jays. We set out bird feeders and grew our first garden on the hillside behind our house there on Cow Creek.

Each landscape demands something from the people living on it.

> By avarice and selfishness, and a grovelling habit, from which none of us is free, of regarding the soil as property, or the means of acquiring property chiefly, the landscape is deformed, husbandry is degraded with us, and the farmer leads the meanest of lives. He knows Nature but as a robber.
>
> —*Walden* (1854) by Henry David Thoreau

Each landscape demands something from the people living on it. It is and always has been difficult to wrest a living from the Appalachian mountains and hills. Subsistence farming may look romantic to an armchair Agrarian, but it is hard, hard work with no guarantees and nothing extra in the best of years. No cash crops. No profit beyond the food you can put up to sustain you through the winter. But a farmer's love and care of the land, though directly based on its utility, transcends it.

Will's family were among those who would not let go of their land. Instead, they used it as well as they could for as long as possible. They were farmers and gardeners in the rich valley of John's Creek. They were storekeepers, teachers, loggers, mineral owners, and leasers. Used to be they used the land, cleared it and planted it, wrested a corn crop, a garden or fruit from it; raised a few hogs; ran some cattle, maybe. Anyone can read in their land the history of a people who knew the value of good soil. Even so, Walker Spears' (Will's grandfather's) farm lay fallow at the mouth of John's Creek from the time he bought it in the mid-1960s until his family began to sell it off nearly thirty years later. It was a rich piece of land that John Walker and Nina McGuire Spears must have been glad to buy with the money from the condemnation of their farm when Dewey Lake flooded the Johns Creek valley. No doubt they envisioned their sons farming it—at least the youngest, maybe, the only one who was still living in Floyd County. That did not come to pass, however. Walker Spears died in gallbladder surgery not long after he'd bought the place, and when I married into the family, the house had rotted down and the only use the land was put to was for an occasional picnic or to hunt and cut our yearly Christmas tree (an unbeautiful yellow pine that signified our connection to this place). When the family finally was ready to let the farm go, in the late 1980s, the buyer wanted it as a place to store heavy equipment, which, of course, meant filling the low bottomland. I still can't look at the ugly, rocky flat where horse weeds once thrived—an indication, so I was told, of rich ground.

Mostly because I needed dirt around my feet, Will and I have lived for the past 26 years on a small farm at East Point. The long valley and low hills of this community on the Floyd/Johnson border looked more like home to me than any other area of the county. We bought an old farmhouse with a red barn and set in to live next to the often-dry bed of George Branch. Imitating Uncle Beta Goble and Uncle Jonah Stepp, we raised gardens (sometimes three in a year) and ran cattle and had goats. In the long shadow of Solomon McGuire, the Johns Creek patriarch of Will's mother's family, we endeavored to keep our hillsides clear enough at least for decent pasture. We watched the backwaters of floods fill the bottom in front of our house, never rising to the level of our yard.

Later we moved up the branch and built a house on a house seat scraped from the hillside and set about improving the fill dirt with compost from our bottomland garden. It was hard to keep the fences up; we learned that firsthand when the cattle and goats kept getting out. We spent a small fortune in fencing. Finally, we sold off our stock. But the hillsides were clear enough, then, from the grazing, to walk, and by the time you'd walked the fence line

in every season or just hiked up to be under those big beeches, the land began to feel like a second skin. Of course, it's still possible to hike the hills, but ours (like most others) have gotten so overgrown that it's not likely we'd go up there. And we are part of the problem, because of changing circumstances, since we no longer use our land for anything (even for gardening) like we used to, but simply "own" it now.

Like most Appalachian people, our family has had a complex and often subtle relationship to their soil, based on its utility as well as the love of this landscape. They've (we've) harvested and husbanded the natural resources that our spot of Earth made available to us. Since these resources did not include much fertile ground, we and our neighbors have looked to the hills themselves for lumber or ginseng to harvest and sell. We've done the best we were able to with whatever mineral wealth the land held. If a man was smart enough and wealthy enough to afford not to sell his minerals (as Solomon McGuire had been), he looked for ways to lease it favorably. Many people (mountain and otherwise) simply have not been that lucky, that wealthy, that foresighted. But we are all complicit in the destruction of our soil—and the mountains.

> The fate of the soil system depends on society's willingness to intervene in the market place, and to forego some of the short-term benefits that accrue from 'mining' the soil so that soil quality and fertility can be maintained over the longer term.
> —Eugene Odum

And so we lend our voices to the protest against the destruction of the once-"everlasting" hills. We particularly protest its more heinous forms, like mountaintop removal. But what do we offer in exchange? What are we willing to give back, to give up? What kind of world do we envision?

116

I wish I had not witnessed the building of US 23's new four-lane incarnation right through the bottomland near our home on George Branch. I would have loved to not wake day after day to the sound of bulldozers and the smell, evening after evening, of newly naked rock faces, as if the earth herself wept her life's blood in the water that seeped from the cuts. I would have loved to still have the bottomland meadow beside the curving two-lane road, its stands of goldenrod and ironweed, the patches of corn some of our neighbors grew, the wide expanse where backwater from the creek and river could collect in seasons of too-much rain. But we don't have any of that anymore.

Taking the easiest and least expensive route, the highway department chose to buy up bottomland in one of the few rural, agricultural neighborhoods left in Floyd county. The East Point Development Club, our community organization and the source of our connection to the community and its history, protested the decision, to no avail. Eventually the club itself was a casualty of the road. What we have now is a 30-foot-high, solid-rock roadbed that blocks run-off, dams up the rush of George Branch after a hard rain, and makes flash floods a real danger. What we have is the sound of cars and trucks 24 hours a day. What we have now is a slightly shorter trip between Prestonsburg and Paintsville. What we have now is a Holiday Inn, several Wal-Marts (both in-use and abandoned), and a downtown filled with pawn shops, mental health clinics, and law offices. Progress—what everyone else has. Though I deplore it, I am also aware that four-lane roads and even the flat land created by mountaintop removal offer room for the kinds of growth that people who live in flat land take for granted.

To be sure, undisturbed mountains alone do not provide jobs, though a beautiful view and the peace of the countryside can feed the soul. But well-cared-for land will nourish us for a thousand years. The point is not to let the hills go back to some prior, idealized, primitive state, or to create in Appalachia a museum of a past way of life—a kind of "reservation" on which the natives live as scenery. We need roads and jobs, and flat land with which to attract both. But more than these, we need to make hard choices that honor the needs of the landscape and its soil equally with the needs of its inhabitants.

When it's all over, I want to say: all my life I was a bride married to amazement. I was the bridegroom, taking the world into my arms.
  —Mary Oliver

What we need is hope that there *is* a sustainable future in the mountains. This means using our knowledge of how ephemeral creeks and bottomlands contribute to a sustainable habitation of the land, reduce flooding, and produce arable soil. It means investment by business, the government, and individuals in small-scale land use—use supported, perhaps, by cooperatives that support local production of food, and by groups like 4-H, local Development Clubs, and the Homemakers clubs that help sustain the culture created when we live in community with each other and the local landscape.

Otherwise we are doomed to live in a decimated place and to participate (if only passively) in its destruction. The present epidemic of drug abuse in Appalachia reminds me of the state of "the projects" in the 1960s and 70s, similarly decimated, overlooked. It is hard to believe that the astounding breadth of drug addiction in Eastern Kentucky does not reflect the ghetto-ization of the Appalachians. After all, to the rest of the world, is this not a place where nothing happens? And according to the prevailing corporate model, aren't the hills of "no use" once the coal in them is gone? The current wave of drug addiction must reflect the hopelessness of Appalachian youth, their feeling of invisibility. It must reflect what it feels like to live in an industrial wasteland where Holiday Inns slide into subsidence holes created by poorly engineered valley fills, and houses are shifted off their foundations by the blasting of surface mines, and the ephemeral stream beds, the sweet clefts where ferns and jewelweed once flourished, are filled with tons of rock and the debris of "progress."

And he called him, and said unto him, How is it that I hear this of thee? Give an account of thy stewardship; for thou mayest be no longer steward.
—Luke 16:2

When There's Just One Left

by Susan Starr Richards

A soft sky pool on my woodland path, magically arranged—
an airy circle, a feathery blue collage, an abstract bird,
a silky still life, with the symmetry of flight. I almost
stoop to touch it, before I let myself know what it is.
Or was. Some noisy jay, I tell myself, then, selective
of my sorrows. But this rust-tinged downy fringe
says bluebird—our lovingly housed radiant domestic sprite.
And why should I discriminate against the jay, anyway?
He makes me laugh, with his right-on red-tail imitation,
giving himself away by just that creaky question mark
at the end, instead of the hard-falling hawkish imprecation,
the raptor's curse: get off my place, you son of a bitch!

But jays make more feathers, and longer ones—
I've come across them, in this condition, right about
here. I look up. Oh yes—that bare ash limb.
The Cooper's hawk sat on its porch, and picnicked
on its smaller avian friend. And even hawks aren't safe,
at night. One spring, on my morning walk, I found both
a Cooper's and a young red-tail—its band still blackish—
each with tails and wings flat to ground, intact,
and in the right places, but with nothing in-between,
as if someone had just painted out, instead of torn out
and carried off, its brains and its soft body. So even
the heavy plunger on rabbits, and the acrobat assassin,
find their great horned nemesis, in the dark. I've heard
that owl sits down beside a roosting bird, crowds it off
its branch, then air-snatches it, with iron claws. And finally
our buzzards trail along, the democrats of death, cruising
the wrinkled wind above this ridge, perusing Earth's leftovers.

So in this bird-eat-bird world, I should wonder why,
in Florida, some years before my birth there, when the last
known breeding pair of ivory-billed woodpeckers appeared,
a taxidermist got permission from the state to cull them,
as too rare to live—shot and stuffed them, to preserve the breed,
and for encased display, that tourists might bring money?

Oh pray for us, all birds, our bloody souls, your unwinged
heirs, that in the hour of our death we go some useful way,
as those two dead hawks I saw surely did go to feed
horned owlets, like the nestful I saw once, chocked
with quaint baby demons, fuzzy white hook-nosed
teddy bears, each following me with its voracious stare;
that we at last may kill only to feed ourselves—
if we had to eat our enemies, maybe we'd get our fill
of war; instead, they're wasted, burnt alive or blown

to bits, unassimilable by our bodies or our minds;
that we might, someday, even learn to practice
Homeland Security as bluebirds do, song by song,
out-trilling instead of out-killing each other.

I dreamed, when I was a child in Florida,
of the ivory-bill—listened for its bass drum
in the flicker's morning rap on our chimney-flashing,
felt its tin trumpet-call and double-knock reign down
in all the teakettle swamps left lost in our suburb,
searched each tree in our back yard, for the long-dead
ever-living Lord-God war-painted scalp-locked
blood-marked primeval bird, on its way
from being one kind of dinosaur to another.

Now it's been found, and we celebrate our old prey,
that we did kill for food, in the beginning, but also for
Indians' war bonnets, then white women's hats;
then for bored tourists; and mainly because
it could fly, and we couldn't, and it was too big
not to bring down. It's been our myth, as one day,
our Kentucky mountains may be. When all our other
mountains have been turned to money—when there's
just one mountain left, will we then learn to love it?
Give it protected status, led to by a grand corridor of
stubbed stumped lopped nubbed 'dozed dynamite-shot
amputated beheaded disfigured nameless creek-choking
unspeakable—don't say plateaus, a plateau is a natural formation,
a mountain melted by millennia, its crown accreted in it,
as in the thirty-foot topsoils of our Bluegrass plateau,
its rich mineralized life rising again in the quick flighty bones
of racehorses. Will we give money to defend our mountain?
Will we put it on a stamp? Presidential decrees, Do Not Disturb
this land? Will children dream of mountains, their deep
heights, here in Kentucky, when the last one's dead?

This bluebird's gone to feathers, a smudge of sky,
here on the mossy floor. But in our diorama world,
one ivory-bill still lives, and it brings tears.

To Sell

fiction by Whitney Baker

Dearest Elena,

I have done it. I am going.

I stood outside his office and craned my neck to his umpteenth floor and shot hate. I wanted to melt the place. I wanted to watch it melt. Right then I was carrying my new mortgage papers. I gave him hell from down there by the parking meters, I reamed him out, I removed his mountaintop, while he was up there chatting at the water cooler, or maybe he was by the window watching me, more certain than usual that I am a lunatic. Then I went up and sold him our land, your land, and I was quiet about it, because he is my brother, because he is your great uncle, because he doesn't know what he doesn't know, like the rest of us idiots.

It was a spectacular try, wasn't it? When it was ugliest, when Esau Wilcox got involved like the old playground all over again, we were all just lost, going around saying God or Lord or Mother Earth when all of us were too shit-faced or shit-on or scared shitless to even face what we were doing. Elena, you know I am a romantic and mouthy to boot; I'm no miner, my hat is soft, and I do put my face down in the loam and good rot and breath of the lady slippers' chaotic castle because I think it is the right and necessary thing to do. Fly a helicopter over me, stagger-hiking that old hill, and I am God's fool, aimless and trying to forget my broken-up life, or remember what good has survived it (you, for one.)

I don't know how fair the deal was. Dean was clean and clinical in there, as always. I could have been Chinese or Martian—it was about "the transaction." Then he was free to ask me about ping-pong and the tomatoes. Some folks on the hill got what seems like more, some less. I don't want to think about it, the dollars of it. I can't let myself get ripped off, of course—we have to pay the hospital, all the "ologists," and whomever else we now owe. I won't spend the rest of my life watching the mountain go blue in my rearview on the drive to my moonlight shift. But I won't count cents as I parse my land into some goddamn assistant pastor's big old gas tank. I got a

122

good enough impression that the price was more or less fair. Lon Campbell says so, and I know he nearly died figuring over his place, mapping it out, wandering over the hills with his laptop and laser compass. I took his opinion and left him with his axe and whetstone and his apartment stinking of rage.

Of course I can't say what your grandmother would have said about all this. I think you are probably right; she would have said down the mountain is better than bankruptcy, better than all-soup-beans-all-the-time and three jobs. I'm more than half-sick of Bell County, of the blue glow from so many windows like space flowers open all night, of the new and new and new Shell stations smearing their fake daylight on the hills. I am sick of the kids at my store so fried they can't remember why they opened their wallets. Now everybody not hauling coal is a patient or a nurse. The nice ones gather up by the creek and watch carp, saying almost nothing, saying I have love enough to come out here and watch these little dragons thrash and be near a person as broken and hopeful as I am.

If you told me you were moving back to these mountains, I'd beg you not to, the same as I'd beg you not to become a therapist for abused children, or a fire jumper. My love for you is a selfish love. There is so much death here just now. I'm no big fan of hope, but I feel sorry for the hopeless in this stained darkness. Elena, I pray I am not one of them, but for now, if you want mountains, go to North Carolina, or Vermont. This place must account for itself. Account for the creepy orange water flowing by the church and for the 13-year-olds vomiting Oxycontin and Ho-Ho's in geography class.

Selling doesn't mean giving up. I'll keep installing the little light bulbs you complain are a little too dull, and I'll let it be a little warmer in summer and cooler in winter than I might. I'll keep talking about it, and keep writing Ernie and Ben, and you. I don't mean to say you are naïve. Youth must be enthusiastic. It is by nature aflame, and I love what you burn for, beautiful girl. Over time you'll mellow, and reconsider, but the deep music you hear now, and the deepest pictures you wake to these mornings, will always be your true guides.

Our little hill will always be gone, like your grandmother. You will have to go there in your mind now; it will be past. Some say that is fine, that the mind is enough, that the Bible is enough. But can I consider the lilies of the field where petite meets electronics at the Krispy Kremes? I don't think so. Their cold world is a lost world, my love. Me, I have to have the menagerie that rots and eats rot to flower. The burning world eats itself, red in tooth and claw, shattered by time but always returning, always seamless, though the hawk tears the warm sparrow full with crickets. And as I walk, I crush legions to see it, wandering to listen, to reconnect, and I am absolutely dumb.

123

Alone in that perfect world, should I talk, I am not understood; if I sit quietly I am lit upon, the sweat-pools on me partaken. Do I think then that I am one with something? I hope I don't think that, or anything. I can't see far enough to know what I might be one with, or what I am a part of. It is foggy here, it is foggy everywhere, and every day is one day less. I don't want to think too much, Elena. I want to know, below my brain and beyond my body. I know this. You are good, and we have done the best that we could do.

Your loving Grandfather

It would be no small advantage if every college were . . . located at the base of a mountain. . . . It were as well to be educated in the shadow of a mountain as in more classical shades. Some will remember, no doubt, not only that they went to the college, but that they went to the mountain.
— Henry David Thoreau

PART THREE

The Law Leaves A Hollow Place

by Amanda Moore

When the last flood tore through Chopping Branch, the water moved Granville Lee Burke's house and washed away his tool shed. The water lingered under his house and ruined the floors. The gravel was already gone from his driveway, swept away by another flood a few weeks earlier. Granville Lee wasn't alone; the floods of the summer of 2002 wreaked havoc all down the hollow, smashing small bridges, lifting pavement from driveways, and damaging several of the old coal-camp houses that sat crowded together between the narrow road and the small stream, both bearing the name Chopping Branch.

I've been told that before the 2002 floods, Chopping Branch, in Letcher County, was not unlike the hundreds of other hollows that snake their way among the steep wooded hills of Eastern Kentucky. It was home to a few dozen families, most of whom were poor or working class, living in old coal-camp houses or mobile homes. People sat on their front porches, watching kids and dogs play in the road that slowly climbed its way up to the Burkes' place. Granville Lee's house, which the county valued at $2,000, sat at the very head of the hollow, sandwiched between his parents' house and his father's garden, where Granville Sr. grew vegetables not for fun but for food. Like many of their neighbors, both Granville Lee and his parents lived on a fixed income received from the government.

In the mid-1990s, the Burkes got a new neighbor. Premier Elkhorn Coal Company began blasting off the tops of the ridges around Chopping Branch to get at the seams of coal running through the mountains. The company took the tons of rock and dirt that it didn't want and dumped them down the side of the mountain towards the head of Chopping Branch. By the time Premier Elkhorn's dynamite and dozers had pushed on past the hills around Chopping Branch, the company had left a 500-foot "valley fill" towering above the hollow. The fill resembled an inverted pyramid with rock-lined ditches running down its sides like tears. At the bottom of the fill was a pond designed to catch the water coming off the huge treeless fill and retain it long enough for the eroded dirt to settle to the bottom. Then, ideally, clear water would trickle out of the pond and into Chopping Branch, serving as a sort of bastardized headwater system for the creek. Granville Lee's house sat just below this sediment pond, separated from it only by his father's garden.

I first learned of Granville Lee's flood damage when I met his father, Granville Sr., at a community meeting near McRoberts, Kentucky, in Letcher County. Several of the residents of Chopping Branch and its neighboring hollows had come together to discuss what to do about Premier Elkhorn's mining operation. It was August 2002, and the recent floods had been the last straw. With the help of a statewide citizens' group, Kentuckians For The Commonwealth (KFTC), the people in the area were organizing, learning their rights under the surface mining laws and strategizing ways to tell the coal company that its abuse of their community had to stop. During the meeting, the KFTC organizer pointed me out as a lawyer with the Appalachian Citizens Law Center. He explained that the Center was located a couple of counties away, in Prestonsburg, and that we provided free legal services to low-income people on coal-mining issues. After the meeting ended, several people, including Granville Sr., approached me and told me about the problems Premier Elkhorn had caused them. I gave them my telephone number and asked them to call me.

Up until that point, my time practicing environmental law in Eastern Kentucky had frequently felt about as productive as banging my head against a brick wall. Week after week, I received calls from people scattered across the coalfields, from Greasy Creek to Mud Creek, from Partridge to Viper, and from dozens of other hollows in between. Even though the people with whom I had spoken were strangers living counties apart, their stories were strikingly similar. They had all believed that a nearby mine had caused some sort of problem, be it blasting damage, flooding, the loss of a water well, or a landslide. Almost always they had filed a complaint with the state agency responsible for regulating mining and had received a letter telling them there was no connection between the problem and the mining. The letter always ended by stating that if the complainant disagreed, he or she could request an administrative hearing in Frankfort, the state capital, by writing a letter within thirty days explaining why the agency was wrong.

The problem, of course, was that the people living in hollows beneath huge mountaintop-removal mines could not afford the hydrologists, engineers, and surveyors that were necessary to show why the agency was wrong. The complainants had always had common-sense explanations as to why they believed the mining had caused their problems— "I've lived here 45 years, and we never had a flood until they cut all the trees off the top of the ridge," or "I felt a huge blast, and within a day, my water turned orange and then disappeared. It'd never done that before." But common sense cannot prevail over advanced degrees in science during an administrative hearing in

Frankfort. I had had to tell several of the people who called our office that unless we could find an expert willing to work for free, we simply could not take their cases, because we would lose.

Of all the calls I had received, the most disturbing were those suggesting a darker force at work within the coalfields and the agencies regulating coal mining. A common refrain among the callers was that someone, either a lawyer or a mine inspector or a neighbor, had been "bought off." I was usually skeptical about individuals receiving actual money to do the coal industry's bidding, but I was well-aware that through its immense political donations and economic importance, the coal industry had essentially "bought off" our entire state government.

The call I received from Granville Sr. a few weeks after the KFTC meeting was different. He told me that as a result of KFTC's organizing efforts, a neighbor's son had learned that surface mining operations were not allowed within 300 feet of an occupied dwelling, and he had thought of the Burkes. The community then got together and made its best effort to measure the distance between the pond at the foot of the valley fill and the two Burke homes. They found that the sediment pond was within the forbidden 300-foot radius from both houses. I was excited to hear this—finally, here was an open-and-shut case, one that required no technical experts to testify. We would simply notify the state agency that Premier Elkhorn was violating the law and then sit back and let the agency do its enforcement work. Case closed, problem solved, no more brick wall.

I was therefore surprised when the agency responded to my letter with a copy of a waiver of the 300-foot requirement that Granville Sr. had signed in May, 1996. When I showed the waiver to Granville, he recognized it, but he said it was a paper the company had given him after it had asked to lease his garden plot and he had refused. "They told me I was just signing something saying I understood that they wouldn't be mining on my property," he said.

Even though, knowingly or not, Granville Sr. had waived his right to the 300-foot protection, Granville Lee had not, and I sent another letter to the agency reminding it of this fact and again requesting immediate enforcement action. Recognizing the potential confusion of having two Granville Burkes owning property side-by-side, I sent the agency a copy of the 1995 deed showing when Granville Sr. and his wife Debra had sold Granville Lee his present house. I also sent the agency a map of the hollow with the Burkes' properties marked on it. Surely, I thought, this will get things moving.

Weeks passed. Nothing happened. By mid-October, I was irritated. I called the agency to speak with the person reviewing Granville Lee's complaint. I was dumbfounded by what I was told: "We don't think Granville Lee really lives in his house, and as you know, the regulation requires that the dwelling be 'occupied.'" "Why in the world," I asked, working hard to keep my voice from getting shrill, "would you not believe that Granville Lee lives in the house that he owns?"

"Well, for starters, there's no telephone line there. Anytime someone wants to call him, they have to reach him by calling his parents' house. And he also shares a post office box with his parents. Plus, the company guys say that when they dug the pond, the house appeared to be uninhabitable. Basically, there's just no indication that he lives on his own in that house."

Again, I did my best not to raise my voice, even though my blood was beginning to thump in my ears. I calmly informed him that the Burkes were far from wealthy and that they shared a telephone line for financial reasons. I then explained to him that the houses in Chopping Branch were mere feet apart and that sharing a telephone number with a relative living next door would not be difficult. "Besides," I said, "why would Granville Lee have bought the house if he didn't intend to live in it?"

The agency employee snorted. "Look, I've bought and sold lots of houses that I never lived in." I couldn't believe what I was hearing. He went on, "And anyway, what took him so long to complain about this? That pond has been there for years."

This time my voice did get a little louder, my tone a little hotter. "Because he didn't know until a couple of months ago that the law, at least in theory, provided him with protection to keep the mining at least 300 feet from his house. He's not a lawyer, you know. He doesn't have the surface mine laws committed to memory."

"Well, let's face it, Ms. Moore," he responded, "your clients aren't exactly the sharpest knives in the drawer."

The conversation ended there. At that point, I realized that the distance between the state government in Frankfort and the people living in places like Chopping Branch is more than just a 185-mile drive. There is a cultural, historic, and economic divide that rural Appalachians spend their lives trying to cross, explain, or deny. This man in Frankfort couldn't understand that in Chopping Branch, it was perfectly normal for a young man to move into the house next to his parents and that circumstances may mandate going without a telephone or living in a house that others see as uninhabitable. He couldn't understand that when a person has had a lifetime of tough breaks, he simply may not question the presence of a valley fill near his home. What's worse, the man's tone, his rudeness about the Burkes, revealed such a deep bias that I wondered how any low-income person in Eastern Kentucky ever got any meaningful assistance from his state government. If a person is viewed as "not the sharpest knife in the drawer," is he therefore not entitled to the protection of the law?

I couldn't stop thinking about my conversation with the man in Frankfort. Not only did I find it professionally frustrating, but I also found myself taking

it a bit personally. As with most of my clients, I genuinely liked the Burkes. I wanted to win some relief for them, not just because that is my job, but because I believed they had gotten a raw deal and deserved some type of restitution.

I had taken on their load to the point where an insult to the Burkes might as well have been a slap to my own face. Within the next few days, Granville Sr. and Granville Lee searched their houses and sent me old utility bills for both homes, Granville Lee's fire insurance policy, and anything else they could find that would show that Granville Lee actually occupied his dwelling. Granville Lee signed an affidavit stating that, yes, he lived in his house. I sent copies of all these things to Frankfort, along with a letter again expressing my strong objection to the burden the agency had placed on my client instead of on Premier Elkhorn.

During this entire period, Premier Elkhorn had been busy buying out or settling with the Burkes' neighbors on Chopping Branch rather than fighting them in court. Immediately after the July floods, the state agency had cited Premier Elkhorn with failing to maintain the sediment pond beside the Burkes' homes. While the agency wouldn't lay all the blame for the floods on Premier Elkhorn, the possibility that a jury might was apparently enough incentive for Premier Elkhorn to settle out of court. The company, however, refused to deal with the Burkes. The Burkes grew frustrated as they watched their community break up, with families either moving away or signing confidential settlements that removed their incentive to organize and fight. As the months passed with no action from the agency and no settlement with the company, the stress of the situation began to weigh on the Burke family.

Finally, in December 2002, the state agency notified Premier Elkhorn that it was breaking the law by placing a sediment pond within 300 feet of Granville Lee Burke's home. The agency gave Premier Elkhorn a choice of how to remedy this violation: either it could reach a deal with Granville Lee that would include his signing a waiver of the 300-foot protection, or it could remove the areas of the pond that fell within 300 feet of Granville Lee's house. The company had thirty days to decide which it would choose.

The likelihood of finally resolving the problems that had started with the floods back in July was welcome news to the Burkes, but the stress of personal and financial problems, combined with living for months with the flood damage, had already taken its toll.

Any hope of a happy ending was shattered when Debra Burke, Granville Sr.'s wife and Granville Lee's mother, took the old gun that had belonged to her father-in-law, and in an upstairs room of their home, shot and killed herself. It was Christmas morning. A little over two weeks later, Premier Elkhorn let the state agency know its decision on remedying the 300-foot violation. Rather than settling with Granville Lee or removing the pond, it appealed the agency's decision and asked for an administrative hearing in Frankfort.

Debra's suicide was devastating; the company's decision to drag out the legal situation with an appeal only added insult to injury. As with most legal proceedings, the process leading up to an administrative hearing in Frankfort is slow. Months passed with no action beyond conference calls with the hearing officer and routine paper filings. All the while, Premier Elkhorn continued to refuse to settle with Granville Lee. Finally, in October 2003, fifteen months after the summer floods had first brought attention to the sediment pond at the head of Chopping Branch, the Burkes and Premier Elkhorn reached an agreement. Afterwards, the Burkes moved across the county, leaving their homes in Chopping Branch behind.

I suppose, in the legal sense, the Burkes eventually won. They made the state agency take enforcement action against Premier Elkhorn, and they got the company to enter into an agreement with them. But what was lost on this path to victory?

Of course the Burke family will never be the same, but the community of Chopping Branch suffered a loss as well. Its residents had united and then divided, split apart by individual deals with the same coal company that had blown up the mountains that had always sheltered Chopping Branch. The lay of the land and the fabric of the community have been changed forever, while Premier Elkhorn's mighty machines, undaunted, continue to devour their way across the ridges of Appalachia, leaving behind a blighted landscape and a litany of tragic stories, like that of the Burke family and Chopping Branch.

All the wild world is beautiful, and it matters
but little where we go, to highlands or
lowlands, woods or plains, on the sea or land
or down among the crystals of waves or high
in a balloon in the sky; through all the
climates, hot or cold, storms and calms,
everywhere and always we are in God's
eternal beauty and love. So universally true
is this, the spot where we chance to be
always seems the best.
        —John Muir

The Long Struggle

by George Brosi and Jerry Hardt

Appalachian coal miners formed the backbone of the union movement for all workers in the 1930s and beyond. Their struggle, in the 70s, for union reform, black lung compensation and mine safety resulted in new precedents which have benefited employees in all fields. But coalfield residents have been just as resolute in attempting to ameliorate the problems of their region as the miners themselves. From the time the practice of strip mining for coal emerged in the 1930s, Eastern Kentucky residents have resisted.

Initially, opposition expressed itself primarily in the form of law suits against the coal companies. In 1943, for example, Ralph and Stella Watson of Magoffin County filed suit to prevent their 20-acre farm from being strip-mined by Elkhorn Coal Company, which held the mineral rights under a broad form deed executed in 1903. The Watsons argued that the sellers of the mineral rights could not have anticipated strip-mining technology, its ability to scrape off the earth down to the coal seam and simply load the coal onto trucks, and thus their forebears did not intend to be selling the right to strip-mine the land. The courts upheld a 1925 ruling and allowed Elkhorn to strip-mine, but initially required that the company pay damages. Two years later, in Pike County, a Russell Fork Coal Company strip mine caused a slide that swept all the homes below down into the valley. The Pike Circuit Court held the company liable but the decision was reversed by the Kentucky Court of Appeals.

Whitesburg lawyer, writer, and state legislator Harry M. Caudill was instrumental in the passage of Kentucky's first bill regulating strip mining in 1954. Chad Montrie, the author of *To Save the Land and People: A History of Opposition to Surface Coal Mining in Appalachia*, termed it "perhaps the weakest of all such legislation at the time," and pointed out that Kentucky was one of the last states to implement regulations. West Virginia, by contrast, has had legislation on the books since 1939.

Responding to Elkhorn Coal's appeal in the previously-mentioned Magoffin County case, that same Court of Appeals, then the highest court in Kentucky, ruled in 1956 that mineral owners could do anything "necessary or convenient" to extract their minerals, and that they were under no obligation to compensate the surface owner for damages unless they were "wanton and

malicious." This important case gave mineral owners unprecedented power over surface owners and pitted landowners against coal companies for the next three decades, in a battle over land rights.

The earliest documented case of direct action against strip miners in Kentucky occurred in 1962, when Warren Wright, a preacher in the Letcher County community of Burdine who would continue to play an important role in opposition to the practice for the rest of his life, blocked the way of Beth-Elkhorn Coal Company's bulldozer while his wife, Mae, sat under a tree holding a loaded pistol in her lap. Wright argued that although his father had sold the mineral rights under a broad form deed, he made sure to include a provision which prohibited the dumping of overburden on his land. Both lower and higher courts ruled against Wright, who responded, "That group of learned reprobates entered an outright lie into the language of their opinion." He later reflected, "That fight made me socially alive for the first time . . . that's when I became a citizen of the United States."

The following year, in 1963, W. D. Bratcher, a Greenville lawyer and leader of the Kentucky League of Sportsmen, organized a meeting at the Game and Fish Commission Office in Frankfort "to investigate the possibility of legislation to control strip mining in Kentucky." He included representatives of the Garden Clubs and Business and Professional Women.

Nevertheless, this group was not consulted by the Farm Bureau and the Conservation Districts when they wrote a new bill which passed both houses of the legislature with only one dissenting vote. A bill written by Harry Caudill to abolish the practice was introduced but ignored by the body.

The threat of strip mining as well as the grass-roots struggle against it in Eastern Kentucky escalated dramatically in 1965. Bill Sturgill, who would later serve on the University of Kentucky Board of Regents and become Governor John Y. Brown's Secretary of Energy, signed a fifteen-year contract to supply the TVA with two million tons of coal stripped from Perry and Knott County mountains and purchased the largest equipment ever used in Kentucky to do the job. Along Clear Creek in Knott County, in the spring of 1965, the company entered land owned by a soldier serving in Vietnam. His stepfather, an eighty-one-year-old coffin-maker and Baptist preacher named Dan Gibson, met them with his rifle and ordered them off the family land. Later, Gibson reflected, "The strip miners are killing these old hills. When they finish, there won't be anything left. Yes, sir. This is my land and my land is dying." *My Land is Dying* became the title of Harry Caudill's 1971 book against strip mining. More than a dozen lawmen arrived that afternoon to arrest Dan Gibson, but he refused to go until he and his supporters had extracted a promise from Sturgill not to strip the land. Then the lawmen took Dan Gibson to the Knott County jail in Hindman. That night, the jail was surrounded by armed men demanding his release. The charges were dropped and he was freed. The next day, a large party of men and women, some armed, met the bulldozers and repulsed them one final time. A few weeks later, on June 1, eighty local strip-mining opponents met in Hindman to organize. The following week an even larger group met and formed the Appalachian Group to Save the Land and People (AGSLP). Leroy Martin, a teacher, was elected Chair for Knott County and Commonwealth Attorney Tolbert Combs for Perry County. Jenkins High School Principal Eldon Davidson was chosen to serve as Chair for Letcher County. At the Group's meeting at Carr Creek School on June 16th, County Judge George Wooten of Leslie County was elected Chair for Leslie County.

Initial Group actions included a fifty-vehicle procession to Frankfort to present a petition with three-thousand signatures to Governor Breathitt. A few days later, Governor Breathitt flew to Knott County and visited the cemetery at the head of Sassafras Creek where Mrs. Bidge Ritchie had watched helplessly in 1959 while strip miners uprooted the coffin of her infant son and pitched it down the mountainside. When Breathitt attended the National Governor's Conference in July 1965, he pushed the governors to take action on the issue. Then, in November, he signed an emergency proclamation to implement new, more stringent regulations of strip mining.

Back in Knott County, Sturgill's Caperton Coal Company was preparing to strip-mine land in Honey Gap owned by Ollie Combs, a sixty-one-year-old widow. She phoned Dan Gibson, who urged her to join the Appalachian Group to Save the Land and People and then brought a posse of armed men to protect her land. The company backed off, but when Gibson's men left a few days later and Mrs. Combs and two of her three triplet sons were there alone, the company returned. Ollie Combs sat down in front of the bulldozer, and Bill Strode, a photographer for the Louisville *Courier-Journal* captured for posterity the image of two law officers carrying her to jail from her property. She and her sons were sentenced to 20 hours in the county jail for violating an injunction Caperton had obtained. Another Strode picture, of Ollie Combs eating Thanksgiving Dinner in jail, also encouraged opposition to strip mining and stimulated Governor Breathitt to declare, "History has sometimes shown that unyielding insistence upon the enforcement of legal rights by the rich and powerful against the humble people of a community is not always the quickest course of action" and to revoke Caperton's mining permit. Legislation introduced to outlaw the broad form deed in the next session of the state legislature became known as the Widow Combs Bill.

Governor Breathitt introduced legislation in the 1966 session essentially designed to institutionalize his emergency regulations to control strip mining. At the hearings, testimony from Judge George Wooten and writer Wendell Berry, as well as from Ollie Combs, Bidge Ritchie and others, proved persuasive. The governor signed the bill at the home of J. O. Matlick, his Conservation Commissioner, who had suffered a heart attack after a confrontation with strip miners. Nevertheless, the legislature tabled the Widow Combs broad form deed bill.

Despite the legislation designed to enforce stricter controls, the number of acres permitted for strip mining increased. So did opposition to the practice. In 1967 hundreds of signatures were presented to U. S. Congressman Carl Perkins of Hindman, and Knott County AGSLP member Mart Shepherd attended a meeting of 75 residents in Harlan County, who formed another chapter of the Group. Before the end of the year, the Harlan Fiscal Court unanimously passed an ordinance opposing strip mining. According to Mary Beth Bingham's article, "Stopping the Bulldozers," in *Fighting Back in Appalachia,* edited by Steven L. Fisher (1993), citizens in the Clear Creek community of Knott County held off the strip miners for months that year. Others worked outside the system. Buck Maggard, a Knott County activist, formed the Mountaintop Gun Club, which volunteered to set up firing ranges on property threatened by strip mining.

Others went further. According to *Our Land, Too,* by Tony Dunbar (1971), over two million dollars in sabotage damage was done to strip-mining equipment in 1967 and 1968.

During the late 1960s, government anti-poverty workers, including Appalachian Volunteers, began to assist local residents opposed to strip mining.

They helped, for example, transport 200 people from eleven counties to a reclamation symposium in Owensboro in 1967. In July of that year, Jink Ray, a farmer in the Island Creek neighborhood of Pike County, organized a large chapter of the Appalachian Group. He and his neighbors turned back the Puritan Coal Company as it attempted to enter his land. Later that month they were part of an AGSLP-organized tour of strip-mining sites in eight counties.

Fifty people participated, including many government officials. Before the end of the year, the mining permit for Jink Ray's land was canceled.

Then on August 11, 1967, Commonwealth Attorney Thomas Ratliff, a candidate for Lieutenant Governor and an officer of a coal association, took aggressive action against the resistance. Accompanied by fifteen armed deputies, he raided the home of Al and Margaret McSurely, employees of the Southern Conference Educational Fund, a non-profit organization that had worked closely with Jink Ray, Appalachian Volunteer Joe Mulloy, and others opposing strip mining. Mulloy and the McSurelys were taken to jail and charged with sedition. In an incident that would have a huge negative impact on the movement against strip mining, Ratliff not only succeeded in leading people to believe that "Communists" were behind the movement; he also identified the movement indelibly with its young supporters who did not have deep roots in the mountain region. Subsequently, some academic historians have insisted that ex-students, not local people, were the leading supporters of the movement. Ironically, of course, it was citizens with deep roots in the mountains, old enough to be the youths' grandparents, who had formed the backbone of the movement from the beginning. Eventually, the McSurelys and Mulloy won a huge settlement from the federal government for violating their rights, but the newspaper headlines connecting "sedition" with the movement against strip mining were much more powerful than the footnote—years later—that officials had violated the rights of the community workers.

In the wake of the raid on the McSurely home, on August 18, 1967, Sergeant Shriver cut off funding for the Appalachian Volunteers. The next blow to the movement is expressed by Harry Caudill in *My Land Is Dying* and reprinted in *Voices from the Mountains*, by Guy and Candie Carawan: "In an opinion handed down on June 21, 1968, by the Kentucky Court of Appeals, the hopes of the Appalachian Group to Save the Land and People and its supporters were crushed. In substance, the majority held that the owner of underlying minerals may totally ruin the surface of the earth without the consent of the man (sic) who owns and tills it—and without paying him anything for his

. . . liberty in a wasteland is meaningless.

loss! Few Kentucky lawyers were surprised by the decision, but among thousands of mountain families it deepened the hopelessness of the old and the cynicism of the young." In the previous December, Governor Breathitt had been replaced by Louie Nunn, a Republican with close ties to the coal industry. This series of setbacks ended the era when the fight against strip mining took place right in the affected areas, when armed local citizens confronted coal company employees on their own land, when a local "culture of resistance" confronted institutional power.

As the Kentucky movement against strip mining lost momentum in 1968, the hope for redress of grievances shifted to the state and national level. In April and May of 1968, the United States Senate Committee on Interior and Insular Affairs held hearings, and Harry Caudill testified, "Let us frankly recognize that the earth is just as important as the people who inhabit it and that the right to be free is matched by a responsibility to preserve freedom's land . . . liberty in a wasteland is meaningless."

In 1969 Luther M. Johnson filed a lawsuit against Bethlehem Steel for violating landowners' rights by devastating the land above them on Millstone Creek in Letcher County. "They are just going to ruin all the land here for the dollar if we don't stop them," he said. "Now I like a dollar as well as anyone. But this land would go on forever if they wouldn't wreck it. And against that, money doesn't seem of much account." On December 12, 1969, the Kentucky Conservation Council met in Whitesburg to consider taking a stronger stand on strip mining, but the results were inconclusive at best.

1971 brought a new governor, Democrat Wendell Ford, to the Statehouse and a new administration to the Council of the Southern Mountains: Warren Wright was elected Executive Director. In June, in front of a packed courtroom, the Knott County Fiscal Court banned strip mining, earning them five minutes of applause. But their action was ignored by the industry and state permit-granters and then, as other county governments began considering similar resolutions, the ban was officially reversed by the state Attorney General who determined that local governments had no authority to regulate mining. Also in 1971, a new organization, Save Our Kentucky, was formed. Joe Begley, of the Letcher County Citizens to Protect Surface Rights, William Cohen, an Alice Lloyd professor, and Warren Wright were involved. The organization hired Jim Branscome, a recent Berea College graduate. Momentum was building slowly, again, at the grass roots. In July a protest was held at the court house in Hindman, and petitions were circulated to revoke strip-mining permits in eight counties. Late in the year, AGSLP activists testified at state hearings, and in January of 1972, 200 citizens traveled to Frankfort to lobby legislators.

That same month, at a Sigmon Brothers mine site on Elijah Fork in Knott County, a group of women, prepared to go to jail in an act of civil disobedience, sat down in the scoop of a huge bulldozer. The women (including Eula Hall, who now runs the Mud Creek Health Clinic in her native Floyd County, and Sally Ward Maggard, a Perry County native who now is a professor at West Virginia University) were ignored by all but press photographers. They stayed fifteen miserable, cold, wet hours; nothing happened until they were forced, by vandalism to their vehicles and physical violence against their male supporters, to leave. Years later, Doris Shepard, who participated in the protest with her mother, Bessie Smith, recalled that the strip-mining controversy divided the community. "Some are still paying for what they did, but it was worth it . . . . We were considered communists, rabble-rousers and hell-raisers," noted Shepherd, saying that mine owners and operators had the protection of the police and politicians. "It was us fighting them—people who were making money. It made us outcasts in our county." People who fought against strip miners were often black-balled from getting jobs. "The strip miners used money to divide families and neighbors," threatening people's pocketbooks. Shepherd explained that her mother, Bessie Smith, got involved after "someone pulled a gun on Grandpa."

She recalled going with her grandpa, when she was a child, to his orchard, a major food source for his family. After their land was stripped, there was no garden, just yellow rocks. "They took his livelihood from him. They had taken his pride," Shepherd said.

After unprecedented floods hit Eastern Kentucky in April 1972, 200 Floyd County residents converged in June to stop a strip-mine operation which many blamed for causing the flooding. That same month, over 900 citizens met at the Union College Environmental Center at Cumberland Gap National Historical Park for a National Conference on Strip-Mining called by U. S. Senator Fred Harris of Oklahoma. Future Kentucky Governor Paul Patton, of Elkhorn Coal, argued that opponents cared only about aesthetic values. Nevertheless, the body as a whole passed a resolution calling for the abolition of strip mining.

The public uproar was getting the attention of Congress. In 1972, the U.S. House passed a bill that banned mining on slopes greater than twenty degrees. A year later, both the U. S. House and Senate held hearings on strip mining. One of the most compelling testifiers was William Worthington, an African-American Harlan County coal miner who had been instrumental in the struggle for black lung compensation.

Just as momentum was beginning to swing toward the side of opponents of strip mining again, everything fizzled out. The oil embargo of the mid-1970s greatly increased the power of the coal industry, and, not surprisingly, marked the beginning of mountaintop removal strip mining, as the West Virginia Department of Environmental Protection permitted 300 acres for this new practice.

Whereas in earlier eras of resistance, local preachers had often led grass-roots struggles, now the clergy of mainstream denominations often took the fight to the halls of power. A powerful pastoral letter, *This Land Is Home to Me,* issued by the 24 Catholic bishops of Appalachia in 1974, spoke of the oppression of the region at the hands of coal: "There is a saying in the region that coal is king. That's not exactly right. The kings are those who control big coal, and all the profit and power which come with it . . . The coal-based industry created many jobs and brought great progress to our country, but it brought other things, too, among them oppression for the mountains."

Church leaders, including the Rev. R. Baldwin Lloyd of the Episcopal Appalachian Peoples Organization and Patrick Ronan of the Catholic Office of Appalachian Ministry in Virginia testified in Congress in support of strong federal controls on mining and an abolition of strip mining: "[T]o hoard, to destroy, or to waste the Earth is to destroy life, and this destruction is wrong and evil," Lloyd testified. "The debate about strip mining is a moral question above all else. Strip mining is immoral because of what it does to people and to land and water and forests, and all other living creatures it affects. . . . There is no wise answer to strip mining but to phase it out as quickly as we can. The moral cost—human and environmental—is too great for it to continue."

In his 1976 book, *Spoil: A Moral Study of the Strip Mining of Coal,* Presbyterian minister Richard Cartwright Austin was blunt in his moral judgment: "The mining is also categorically evil in the strictest moral sense; not only is it enormously destructive of men and the earth, but every single element of its destructiveness is unnecessary and preventable. It is now sin— the perversion of good into evil—for the destructiveness is not prevented."

In April 1977, another round of devastating floods occurred in Eastern Kentucky, the worst ever experienced downstream from strip-mining operations. Communities such as Cranks Creek experienced three devastating floods in 1977, coming with less and less rainfall, because the creek had filled so much with silt from the strip mines. In Martin County, citizens formed the Concerned Citizens of Martin County when the Martin County Housing Authority unsuccessfully tried to use a Federal Community Development Block Grant to relocate the town of Beauty to clear the way for strip mining. In Harlan County, citizens formed both the Cloverfork Organization to Protect the Environment (COPE) and the Cranks Creek Survival Center in response to the flooding and ongoing problems with strip-mining abuses.

In the 1970s, three attempts by Congress to pass a strip-mining law were defeated by vetoes, or threatened vetoes, by Presidents Richard Nixon and Gerald Ford.

In the summer of 1977, with Democrat Jimmy Carter in the White House, Congress again considered legislation. In testimony before the U.S. House Subcommittee on Energy and the Environment of the Committee on Interior and Insular Affairs, Paul Patton testified in opposition to key provisions of the bill. He called many provisions of the bill "excessive" and specifically spoke against the provision in the law that stripped land be returned to the "approximate original contour," a requirement he failed to enforce as governor and which still eludes effective enforcement. Patton did admit, in his testimony, however, "I don't contend that all strip-mining ventures—or half of them or a big percentage of them—will ever be utilized for a higher land use."

In the same hearing, Robert Bell, commissioner of the Kentucky Department of Natural Resources, testified, "Placement of large volumes of uncompacted spoil on steep slopes below the elevation of the coal seam will generally result in landslides and severe erosion, and debris. The proposed act prohibits placement of overburden over the out slopes. This is a strong provision, and Kentucky supports this provision." Obviously, he did not envision the massive valley fills that accompany mountaintop removal operations today. In the legislative process the ban against steep slope mining was removed from the bill.

President Jimmy Carter signed the first federal legislation regulating, but not banning, strip mining in an August 1977 ceremony attended by about 300 guests in the Rose Garden of the White House. Some activists had urged President Carter to veto the legislation on the grounds that it was not strong enough and would not be inforced.

In 1978, in response to a complaint by Hazel King of COPE, the new Federal Office of Surface Mining issued the first cease-and-desist order against the Easton Deaton mine on Clover Fork in Harlan County. Since then, residents and groups across the coalfields have fought for the law's full enforcement—a significant change from previous years, when the focus was on stopping strip mining altogether.

In December of 1981, a broad coalition of groups known as the Appalachian Alliance, formed at a meeting on the Union College campus in 1977 in response to the floods and their aftermath, hired Joe Szakos as field organizer for Kentucky. On August 17, 1981, people from twelve counties met and formed the Kentucky Fair Tax Coalition (KFTC). The organization changed its name to Kentuckians For The Commonwealth (still KFTC) on January 1, 1988. Though initially focusing on other coal issues, KFTC's focus quickly formed around the effort to end the abuses of the broad form deed that allowed coal companies to strip-mine without the permission of the landowners. The Widow Combs Bill that had failed in the late 1960s was eventually passed by the General Assembly in 1974, in an effort led by Representative Raymond Overstreet. But that law was ruled unconstitutional by the Kentucky Court of Appeals. Renewed efforts by Overstreet and others in subsequent legislative sessions were stymied by legislative leaders with ties to the coal industry.

With growing public discontent over this blatant injustice, and with some new legislative faces from Eastern Kentucky, KFTC introduced a bill to restrict mineral-rights owners from mining in ways not envisioned when the original broad form deeds were purchased (generally between 1880-1910). Over 200 supporters rallied in Frankfort to support the legislation. It passed, and Governor Martha Layne Collins signed it. It took lawsuits from Everett Akers of Floyd County (a former legislator who had been arrested for trespassing on his own property under the broad form deed), Joe Begley of Letcher County, Joe Whitaker of Leslie County, Elizabeth Wooten of Perry County, and Chalmer and Marie Hicks of Knott County to get the permitting authorities to recognize the new law, but the Kentucky Court of Appeals ruled in favor of the citizens.

In the summer of 1986, Pike County citizens living on Long Fork protested Cobra Coal Company's application for a permit to strip-mine. The County Property Tax Assessor's son worked for that company, and everyone who testified against permitting the coal company had their property assessment raised. They confronted the Tax Assessor, and he backed down.

To mark the tenth anniversary of federal legislation on strip mining, coalfield activists from around the country met in 1987 in Lexington, Kentucky, and formed the Citizen's Coal Summit, which has evolved into the Citizens' Coal Council and remains an important coalition to this day.

In July 1987, in a confusing and split decision, the Kentucky Supreme Court ruled the Kentucky Broad Form Deed Law unconstitutional. Immediately, KFTC initiated a campaign to amend the Kentucky

Constitution to provide a legal interpretation of the broad form deed that would prohibit strip mining without the permission of the surface owner. Representative N. Clayton Little of Pike County and Senator Benny Ray Bailey of Knott County led the effort, which resulted in both houses of the legislature unanimously agreeing to put the issue on the ballot.

A popular vote was set for November 2, 1988. With KFTC leading the grass-roots statewide campaign, the amendment won, garnering an astonishing 82.5 percent of the vote statewide, with the largest margins coming in the eastern coal counties. This ushered in a new era in the struggle against strip mining. Now, strip mining could not take place without the permission of the landowner. The only people who could object were adversely affected adjoining landowners and those from further away.

At the end of 1988, as he was leaving office, in order to protect the "rights" of mineral owners, President Ronald Reagan proposed doing away with the buffer zones established in the 1977 law to protect adjacent homes, schools, cemeteries, and national parks from strip-mining. OSM sponsored hearings on these initiatives. Citizens from across the Kentucky and Appalachian coalfields testified against these proposals, leading to their withdrawal.

In 1989, over 200 citizens of the Woodbine area of Knox County initiated action to force the Federal Abandoned Mine Lands program to reclaim a strip mine abandoned without reclamation by the Richlands Coal Company. They succeeded not only in getting a new water system for their community, but also in getting the local mine inspector, whom they accused of being lax on Richlands, transferred.

In July 1993 the Kentucky Supreme Court upheld the Broad Form Deed Amendment against a challenge by the Lash Coal Company. John Rosenberg, of the Appalachian Research and Defense Fund, presented the winning case, effectively ending the era of the broad form deed, but not the impact of strip mining on land adjacent to vast holdings owned or leased by coal companies.

The amount of stripped acreage continues to grow. The ruling also did nothing to stop the regulatory agencies from favoring the coal companies over the afflicted citizens. That same year, a Congressional investigation sharply criticized OSM for failure to enforce properly the federal strip-mine law, including inappropriate actions by then-OSM Director Harry Snyder to help coal operators. Snyder sometimes called field agents after work hours with instructions contrary to policy and regulations. The report documented the systematic dilution of the federal mining law since 1980 by the U.S. Interior Department through the weakening of rules and regulations, and also through reorganizations that "rendered OSM mine inspections ineffectual." Reclamation inspectors accounted for only nine percent of the agency's one thousand employees, and received only six percent of the agency's $270 million annual budget.

In 1995, residents of Long Fork in Pike County organized a powerful alliance among truck drivers and community members to win commitments from the Premier Elkhorn Coal Company to address local concerns that included coal dust, unsafe roads and bridges, the fear of economic blackmail against truckers and citizens who spoke out, good mining jobs not going to local people, and unsafe silt ponds. Local residents supported a successful strike by Long Fork truckers to oppose an attempt by Premier to make the coal truck drivers take an unsafe route and accept a cut in pay. Their demands were met, but nine drivers were immediately laid off, while the company hired two additional trucks that were not from the community. A few days later, more than fifty people attended a community meeting. The nine truckers were called back to work at midnight, just hours after the community meeting ended.

In 1996, more than 100 Pike Countians from the Feds Creek and Mouth Card areas attended a public meeting to support the extension of public water lines to areas where years of mining had destroyed their water wells. Pike

County school children from the area went to a Fiscal Court meeting and offered free lemonade made with water from their school. There were no takers. The construction of a promised new school to replace their current building, condemned since the 1960s, was being delayed because there was no longer a reliable water supply in the area.

The loss of water was a concern throughout the coalfields. Thousands of families have seen their wells damaged or destroyed by mining. Residents of eastern Perry County, where several coal companies have contributed to a sprawling 6,200-acre mountaintop removal operation, have demonstrated in Hazard and Ashland (headquarters of one of the companies) and held numerous community meetings and accountability sessions with officials. An informal survey conducted in the community identified nearly 750 households that had experienced water problems associated with mining in the area.

Getting the coal companies to obey the law and enforcement agencies to enforce it has been at the heart of most of these local efforts.

Harlan County residents learned in the late 1990s of Jericol Mining Company's plans to strip-mine near the peak of Black Mountain, the tallest mountain in Kentucky. Black Mountain is 500 feet higher than neighboring peaks in Kentucky, creating a biologically unique area that supports at least fifty plants and animals found nowhere else in Kentucky. Judy Hensley decided that studying Black Mountain and the issues involved would be a good science project for her seventh-grade class at nearby Wallins Elementary School. That effort snowballed into a campaign to save the mountain that also involved students from Evarts High School in Harlan County, and Rosenwald-Dunbar Elementary in Jessamine County. The students testified about Black Mountain before a legislative committee in December 1998. Saving Black Mountain became a campaign that media statewide picked up. KFTC filed a rarely used "lands unsuitable for mining" petition to have Black Mountain declared off-limits to strip mining. Support grew around the state. In 1999, with public sentiment turning against them, and encouraged by the direct intervention of Governor Paul Patton, the eight corporate land and mineral owners with an interest on Black Mountain signed an agreement with KFTC that kept the upper elevations of Black Mountain safe from mining and logging. The agreement called for the state to purchase the coal and timber rights, and the following year the General Assembly appropriated the money to do this.

In 2000, the Board of the Pine Mountain Settlement School learned that Nally & Hamilton wanted to mine near its campus in Harlan County. The school, built in 1913 and designated an historic site in 1991, provides environmental education and an outdoor classroom to about four thousand students a year. With the help of the Kentucky Resources Council, the settlement school also filed a petition for a "lands unsuitable for mining" designation to protect 5,266 acres. A public hearing in February 2001 lasted seven hours over the span of two evenings with dozens speaking. The state received more than 2,300 letters about the petition; only 76 of those letters opposed the petition. In April 2001, the Kentucky Natural Resources Secretary James Bickford declared 2,364 acres surrounding the Pine Mountain Settlement School unsuitable for strip mining.

One of the key battles throughout the 1990s and into the present has been against ongoing administrative attempts to weaken or ignore existing laws. For instance, in 2002 the Bush Administration changed a key definition in the Clean Water Act to allow mining wastes to be placed in or near streams. The change in definition involved the distinction between "waste" and "fill material." The Clean Water Act prohibits "waste" being dumped in streams. They had broken the law before this change in definition, but now the mountaintop-removal operations could continue the same practice, legally pushing the overburden (formerly waste) over the side of the mountain, creating "valley fills" that have buried nearly 500 miles of Kentucky streams. The Bush action came in response to an August 2001 lawsuit by KFTC against the U.S. Army Corps of Engineers for issuing valley-fill permits to Martin County Coal to create 27 valley fills that would bury nearly six miles of streams. The Corps considered the valley fills to have a negligible environmental impact.

In May 2002, U.S. District Judge Charles Haden II, a conservative Republican, struck down the administration's new rule in a decision that read in part, "Amendments to the act should be considered and accomplished in the sunlight of open congressional debate and resolution, not within the murky administrative after-the-fact ratification of questionable regulatory practices. The agencies' attempt to legalize their longstanding illegal regulatory practice must fail." However, the U.S. Appeals Court in Richmond, Virginia, negated Haden's decision and validated the new regulations or interpretation of regulations.

In May and June 2002, Kentuckians joined dozens of coalfield residents from other states in Washington, D.C., to lobby Congress in support of legislation to overturn Bush's Clean Water Act rule change. The June trip included a subcommittee hearing of the Senate Environment and Public Works Committee focused on legislation known as the Shays-Pallone Act, to restore the original definition of fill material into the Clean Water Act so that mountaintop-removal mining could not simply push the overburden into waterways. Republican members boycotted that meeting, and the bill has not been allowed a hearing by current House and Senate leaders.

Churches and church leaders, speaking regularly at rallies and events to support the protection of the land and people of the coalfields, have remained important players in more recent resistance to mountaintop removal. In December 2002 and May 2003, the Catholic Diocesan Office for Justice and Peace in Lexington and KFTC organized two Prayer on the Mountain services hosted by the Reverend Steve Peake at the Corinth Baptist Church in Neon. These were endorsed by the Kentucky Council of Churches and the Commission on Religion in Appalachia (CORA). CORA is among several official church bodies that passed resolutions in the mid- to late-1990s calling for an end to mountaintop removal.

More recent demonstrations and statements and organizing efforts have called for the abolition of mountaintop removal and valley fills. Drastically declining employment in the mines, the erosion of mining wages through union-busting, and the monstrous level of environmental destruction have dwarfed whatever benefits the people of the region might once have realized from coal mining.

In June 2003, more than 500 people attended a "Flyover Festival" at the Hazard airport to get a firsthand look at mountaintop removal and support a campaign calling for its abolition. Organized by KFTC and originally intended as an opportunity for about forty KFTC leaders from across the state to view the mining, the event captured the spirit of a growing, broad-based movement to abolish the practice. Two weeks after the festival, more than 125 people supported the call for abolition at a rally in Lexington. In April 2005 a flyover was part of the Kentucky Author's Mountaintop Removal Tour, initiated by Wendell Berry. The event also included listening to testimony of local residents and visiting both undisturbed mountain terrain and mining sites by car and by foot. A resulting author's statement, drafted by novelist Silas House, who comes from a family of Eastern Kentucky coal miners, supported abolition of mountaintop removal. In June 2005, 175 people participated in a Lexington rally to abolish mountaintop removal, which was organized by Mountain Justice Summer.

Current litigation brought by the Kentucky Resources Council, Appalachian Citizens Law Center, Kentucky Waterways Alliance, Kentucky Riverkeeper, the Floyds Fork Environmental Association, and KFTC challenges the process by which the Corps rubber-stamps valley-fill permits without any type of environmental assessment or public review. Another lawsuit is challenging the failure of the state and the U.S. Environmental Protection Agency to enforce an anti-degradation provision of the Clean Water Act that would help protect coalfield streams.

In the courts, in the halls of government, in the churches and schools, and on the slopes, Kentuckians have refused to stand by passively as their mountains have been turned into disaster areas. Those now seeking to protect the land and people are part of a generations-long struggle. Important victories have been won through persistence and discipline, but the final triumph will come only with even greater tenacity and vision.

Information on how to contact your federal, state, and
local officials can be found on the internet at
http://www.congress.org/

The Ballad of Dan Gibson

by Gurney Norman

The foreman of the 'dozer crew
Told old Dan he was coming through
To scrape away his trees and earth
And strip-mine coal for all he was worth.

Dan looked at the 'dozer and then at the fellow,
Said "Buddy you're wrong if you think I'm yellow.
Don't think that because I'm an old man
I'm afraid to defend my piece of land."

The driver of the 'dozer had a second thought.
He began to wonder just whether he ought
To tangle with as fierce-a-looking man
As the one before him they called old Dan.

So he said to Dan, "Now listen, mister,
I drive this 'dozer for my wife and sister.
I ain't drawing a big enough pay
To die for no boss and the TVA.

"So let me tell you what I'm going to do,
I'll go and get the boss of this strip-mine crew.
Then if you two are bound to fuss
Why that'll be better than the two of us."

The foreman went to Hazard and got the boss,
Told him he was scared their cause was lost.
"If we don't do something about old Dan,
We'll never get to strip-mine any more land."

The stripper could hardly believe his ears.
For how could a man of eighty years
Defy an industry as powerful as coal?
The stripper couldn't see it to save his soul.

Old Dan wasn't hard for them to find.
The cops could have found him if they'd been blind.
For Dan wasn't the kind to cringe and hide,
He figured the law was on his side.

But the boss said, "Dan, that's where you're wrong.
The law's on the side of him that's strong.
That's why us coal men are all so lucky
To mine in the Commonwealth of Kentucky."

Dan could hardly believe his ears.
His bright old eyes filled up with tears.
He wasn't afraid to go to jail,
But he couldn't stand to think that freedom had failed.

The strip-miner said. "Now don't cry, Dan.
Let's make us a deal for this here land.
If you let me strip this worthless hill,
Me and my partners will pay you well."

Dan looked around him at all the cops.
Then he gazed awhile at the mountain tops.
He saw the ugly strip-mine bench,
Then frowned at the boss like he smelled a stench.

Said, "You got this coal through a crooked deed
To satisfy your lust and greed.
You cheated my family of the mineral rights
But you'll never get the surface without a hard fight."

So they took old Dan to the Hindman jail.
But he got out on a sudden bail
Put up by his neighbors and all of his kin
So he could fight the strip-miners again.

Dan's fighting them now, with a thousand others
Who in this war are all blood-brothers,
Bound to defend each other's land
From the ruin of the greedy strip-mine band.

No matter how sophisticated you may be, a
large mountain cannot be denied—it speaks
to the very core of our being.
                    —Ansel Adams

Compromise, Hell!

by Wendell Berry

We are destroying our country—I mean our country itself, our land. This is a terrible thing to know, but it is not a reason for despair unless we decide to continue the destruction. If we decide to continue the destruction, that will not be because we have no other choice. This destruction is not necessary. It is not inevitable, except that by our submissiveness we make it so.

We Americans are not usually thought to be a submissive people, but of course we are. Why else would we allow our country to be destroyed? Why else would we be rewarding its destroyers? Why else would we all—by proxies we have given to greedy corporations and corrupt politicians—be participating in its destruction? Most of us are still too sane to piss in our own cistern, but we allow others to do so, and we reward them for it. We reward them so well, in fact, that those who piss in our cistern are wealthier than the rest of us.

How do we submit? By not being radical enough. Or by not being thorough enough, which is the same thing.

Since the beginning of the conservation effort in our country, conservationists have too often believed that we could protect the land without protecting the people. This has begun to change, but for a while yet we will have to reckon with the old assumption that we can preserve the natural world by protecting wilderness areas while we neglect or destroy the economic landscapes—the farms and ranches and working forests—and the people who use them. That assumption is understandable in view of the worsening threats to wilderness areas, but it is wrong. If conservationists hope to save the wild lands and wild creatures, they are going to have to address issues of economy, which is to say issues of the health of the landscapes and the towns and cities where we do our work, and the quality of that work, and the well-being of the people who do the work.

Governments seem to be making the opposite choice, believing that the people can be adequately protected without protecting the land. And here I am not talking about parties or party doctrines, but about the dominant political assumption. Sooner or later, governments will have to recognize that if the land does not prosper, nothing else can prosper for very long. We can have no industry or trade or wealth or security if we don't uphold the health of the land and the people and the people's work.

It is merely a fact that the land, here and everywhere, is suffering. We have the "dead zone" in the Gulf of Mexico and undrinkable water to attest to the toxicity of our agriculture. We know that we are carelessly and wastefully logging our forests. We know that soil erosion, air and water pollution, urban sprawl, the proliferation of highways and garbage, are making our lives always less pleasant, less healthful, less sustainable, and our dwelling places more ugly.

Nearly forty years ago my state of Kentucky, like other coal-producing states, began an effort to regulate strip mining. While that effort has continued, and has imposed certain requirements of "reclamation," strip mining has become steadily more destructive of the land and the land's future. We are now permitting the destruction of entire mountains and entire watersheds. No war, so far, has done such extensive or such permanent damage. If we know that coal is an exhaustible resource, whereas the forests over it are with proper use inexhaustible, and that strip mining destroys the forest virtually forever, how can we permit this destruction? If we honor at all that fragile creature, the topsoil, so long in the making, so miraculously made, so indispensable to all life, how can we destroy it? If we believe, as so many of us profess to do, that the Earth is God's property and is full of His glory, how can we do harm to any part of it?

If we believe, as so many of us profess to do, that the Earth is God's property and is full of His glory, how can we do harm to any part of it?

In Kentucky, as in other unfortunate states, and again at great public cost, we have allowed—in fact we have officially encouraged—the establishment of the confined-animal-feeding industry, which exploits and abuses everything involved: the land, its people, the animals, and the consumers. If we love our country, as so many of us profess to do, how can we so desecrate it?

But the economic damage is not confined just to our farms and forests. For the sake of "job creation" in Kentucky, as in other backward states, we have lavished public money on corporations that come in and stay only so long as they can exploit people more cheaply here than elsewhere. The general purpose of the present economy is to exploit, not to foster or conserve.

Look carefully, if you doubt me, at the centers of the larger towns in virtually every part of our country. You will find that they are economically dead or dying. Good buildings that used to house needful, useful, locally-owned small businesses of all kinds are now empty or have evolved into junk stores or antique shops. But look at the houses, the churches, the commercial buildings, the courthouses, and you will see that more often than not they are

156

comely and well-made. And then go look at the corporate outskirts: the chain stores, the fast-food joints, the food-and-fuel stores that no longer can be called service stations, the motels. Try to find something comely or well-made there.

What is the difference? The difference is that the old town centers were built by people who were proud of their place and who realized a particular value in living there. The old buildings look good because they were built by people who respected themselves and wanted the respect of their neighbors. The corporate outskirts, on the contrary, were built by people who manifestly take no pride in the place, see no value in lives lived there, and recognize no neighbors. The only value they see is the money that can be siphoned out to more fortunate places—that is, to the wealthier suburbs of the larger cities.

Can we actually suppose that we are wasting, polluting, and making ugly this beautiful land for the sake of patriotism and the love of God? Perhaps some of us would like to think so, but in fact this destruction is taking place because we have allowed ourselves to believe, and to live, a mated pair of economic lies: that nothing has a value that is not assigned to it by the market, and that the economic life of our communities can safely be handed over to huge corporations.

We citizens have a large responsibility for our delusion and our destructiveness, and I don't want to minimize that. But I don't want to minimize, either, the large responsibility that is borne by government.

It is commonly understood that governments are instituted to provide certain protections that citizens individually cannot provide for themselves. But governments have tended to assume that this responsibility can be fulfilled mainly by the police and the military. They have used their regulatory powers reluctantly and often poorly. Our governments have only occasionally recognized the need of land and people to be protected against economic violence. It is true that economic violence is not always as swift, and is rarely as bloody, as the violence of war, but it can be devastating nonetheless. Acts of economic aggression can destroy a landscape or a community or the center of a town or city, and they routinely do so.

Such damage is justified by its corporate perpetrators and their political abettors in the name of the "free market" and "free enterprise," but this is a freedom that makes greed the dominant economic virtue, and it destroys the freedom of other people along with their communities and livelihoods. There are such things as economic weapons of massive destruction. We have allowed them to be used against us, not just by public submission and regulatory malfeasance, but also by public subsidies, incentives, and sufferances impossible to justify.

We have failed to acknowledge this threat and to act in our own defense. As a result, our once-beautiful and bountiful countryside has long been a colony of the coal, timber, and agribusiness corporations, yielding an immense wealth of energy and raw materials at an immense cost to our land and our land's people. Because of that failure, also, our towns and cities have been gutted by the likes of Wal-Mart, which have had the permitted luxury of destroying locally-owned small businesses by means of volume discounts.

Because as individuals or even as communities we cannot protect ourselves against these aggressions, we need our state and national governments to protect us. As the poor deserve as much justice from our courts as the rich, so the small farmer and the small merchant deserve the same economic justice, the same freedom in the market, as big farmers and chain stores. They should not suffer ruin merely because their rich competitors can afford (for a while) to undersell them.

Furthermore, to permit the smaller enterprises always to be ruined by false advantages, either at home or in the global economy, is ultimately to destroy local, regional, and even national capabilities of producing vital supplies such as food and textiles. It is impossible to understand, let alone justify, a government's willingness to allow the human sources of necessary goods to be destroyed by the "freedom" of this corporate anarchy. It is equally impossible to understand how a government can permit, and even subsidize, the destruction of the land and the land's productivity. Somehow, we have lost or discarded any controlling sense of the interdependence of the

Earth and the human capacity to use it well. The governmental obligation to protect these economic resources, inseparably human and natural, is the same as the obligation to protect us from hunger or from foreign invaders. In result, there is no difference between a domestic threat to the sources of our life and a foreign one.

It appears that we have fallen into the habit of compromising on issues that should not, and in fact cannot, be compromised. I have an idea that a large number of us, including even a large number of politicians, believe that it is wrong to destroy the Earth. But we have powerful political opponents who insist that an Earth-destroying economy is justified by freedom and profit. And so we compromise by agreeing to permit the destruction only of parts of the Earth, or to permit the Earth to be destroyed a little at a time—like the famous three-legged pig that was too well-loved to be eaten all at once.

Our destructiveness has not been, and it is not, inevitable. People who use that excuse are morally incompetent, they are cowardly, and they are lazy.

The logic of this sort of compromising is clear, and it is clearly fatal. If we continue to be economically dependent on destroying parts of the Earth, then eventually we will destroy it all.

So long a complaint accumulates a debt to hope, and I would like to end with hope. To do so, I need only repeat something I said at the beginning: Our destructiveness has not been, and it is not, inevitable. People who use that excuse are morally incompetent, they are cowardly, and they are lazy. Humans don't have to live by destroying the sources of their life. People can change; they can learn to do better. All of us, regardless of party, can be moved by love of our land to rise above the greed of our land's exploiters. This, of course, leads to practical problems, and I will offer a short list of practical suggestions.

We have got to learn to respect ourselves and our dwelling places. We need to quit thinking of rural America as a colony. Too much of the economic history of our land has been that of the export of fuel, food, and raw materials that have been destructively and too-cheaply produced. We must reaffirm the economic value of good stewardship and good work. For that, we will need better accounting than we have had so far.

We need to reconsider the idea of solving our economic problems by "bringing in industry." Every state government appears to be scheming to lure in a large corporation from somewhere else by "tax incentives" and other squanderings of the people's money. We ought to suspend that practice until we are sure that in every state we have made the most and the best of what is already there. We need to build the local economies of our communities and regions by adding value to local products and marketing them locally before we seek markets elsewhere.

We need to confront honestly the issue of scale. Bigness has a charm and a drama that is seductive, especially to politicians and financiers; but bigness promotes greed, indifference, and damage, and often bigness is not necessary. You may need a large corporation to run an airline or to manufacture cars, but you don't need a large corporation to raise a chicken or a hog. You don't need a large corporation to process local food or local timber and to market it locally.

And, finally, we need to give an absolute priority to caring well for our land—for every bit of it. There should be no compromise with the destruction of the land or of anything else that we cannot replace. We have been too tolerant of politicians who, entrusted with our country's defense, become the agents of our country's destroyers, compromising on its ruin.

And so I will end this by quoting my fellow Kentuckian, a great patriot and an indomitable foe of strip mining, the late Joe Begley of Blackey: "Compromise, hell!"

Spring in Kentucky hills will soon awaken;
The sap will run in every vein of tree . . .
Spring in Kentucky hills and I shall be
A free-soil man to walk beneath the trees
And listen to the wind among the leaves
And count the stars and do as I damn please.
　　　　　　　　　　　　　　—Jesse Stuart

Rate Too Low

by Erik Reece

I spent the last two years writing a book about strip-mining in Eastern Kentucky. It has been an education. I've seen trailers so damaged by mine blasting that you can't set a glass on the kitchen table without it sliding off.

I've seen miles of streams buried under the spoil created by mountaintop-removal mining, and I've seen miles of streams running acid-orange from mine drainage.

I've seen whole communities nearly washed away by flash floods that pour down valley-fills like water rushing through a funnel.

I've seen homes that sit so close to unreclaimed hollow fills that a strong storm and a mudslide could instantly bury all who live there.

I've been to Martin County and seen the sludge that still lines the bottom of Coldwater Creek four years after a coal slurry impoundment broke and flooded the valley with its toxic black waves.

It is not news that people in the coalfields have problems that are the result of an extractive industry that has taken much from the region and given little back. Governor Ernie Fletcher says he will push for long-overdue tax modernization in Kentucky. As the Kentucky General Assembly reconvenes, I urge its members to do something very basic and very substantive: raise the coal-severance tax and redirect more of that money to the counties most affected by strip-mining.

The tax, established in 1976, charged 4.5 percent on each ton of mined coal. Almost thirty years later, that rate hasn't changed, but the damage caused by mountaintop removal has increased drastically. Wyoming and West Virginia, the only states that mine more coal than Kentucky, have severance taxes set at seven percent and five percent, respectively. I urge the state legislature to split that difference and raise Kentucky's coal severance tax to six percent.

Recently, the price of coal jumped from $34 to $55 per ton. The companies can certainly afford a modest increase in the severance tax. More than half of the tax now goes into a general fund to be spent on projects such as the renovation of Lexington's civic center. This is simply unfair. The money should go to the mountain regions that produce it.

While the coal industry likes to tout its ability to create jobs in Eastern Kentucky, the poorest counties in Kentucky are the ones that have seen the most strip-mining. Therefore, a much higher percentage of the coal-severance tax should be returned to those counties to repair homes and streams, and to create good, sustainable jobs.

Over the last two decades, $70 million of the severance tax has gone toward building eight industrial parks across Eastern Kentucky. These have not been a success. The Coalfields Industrial Park in Hazard is almost empty. And $8.5 million was spent on the Bluegrass Crossing Regional Industrial Park, which today has one tenant that employs 35 people making air bags. Unless each of those employees is making $245,000 a year and pouring it back into the community, that was not money well spent.

Multinational companies have come to these industrial parks, then moved on to places like El Salvador, where labor is even cheaper than in Eastern Kentucky. If globalization has not brought prosperity to the mountains, perhaps it is time to consider its opposite: a regional economy. I wonder how different Eastern Kentucky would look today if those subsidies had gone to local people, local businesses, and local cultural institutions.

The mountains of Kentucky are not bereft of tradition—far from it. Just listen to the title track of Loretta Lynn's great new album, *Van Lear Rose*, to understand how art can come from hard times and can give them meaning and beauty.

But since John C.C. Mayo began buying up mineral rights in the 1880s, the mountains and the people of Eastern Kentucky have been exploited and treated poorly so that the rest of us could have cheap electricity.

It is time the coal industry and the rest of Kentucky understood that sacrifice and started paying back a debt that is long overdue.

I'm proud to be a coal miner's daughter.
　　　　—Loretta Lynn

All That We Have
— a song to sing

by Anne Shelby

I was born in these mountains
Lived here all my life
And I thank the good Lord
Every day I'm alive
For these tree-covered hillsides
And clear rushing streams
That come down off the mountain
And run through my dreams

My family, they've been here
For two hundred years
Watered these hillsides
With their sweat and their tears
Working together
Working this land
For this land and each other
Was all that they had

Don't tear down these mountains
With your mighty machines
Don't fill up these valleys
And poison these streams
Don't bury the topsoil
And destroy the good land
For the land and each other
Are all that we have

My folks are all gone now
I'm left here alone
But their sweet gentle spirits
Are with me, I know
They linger on the ridge-tops
And in the cool coves
A part of this parcel
Of Earth that they loved

Now the neighbors are selling
And moving away
And the roar of the dozers
Gets closer each day
They're felling the forests
They're leveling the hills
For a few dollars off
Electricity bills

Don't tear down these mountains
With your mighty machines
Don't fill up these valleys
And poison these streams
Don't bury the topsoil
And destroy the good land
For the land and each other
Are all that we have

Oh, the rich they get richer
And the poor mine the coal
And the lights must keep burning
In the cities, we're told
But where will we turn to
When the boom turns to bust
And the once-verdant mountains
To rock piles and dust

We are living in the last days
Some people say
Look for Christ's Second Coming
And the Great Judgment Day
But what will the Lord say
When he sees what we've done
With this Garden of Eden
He gave us for a home

Don't tear down these mountains
With your mighty machines
Don't fill up these valleys
And poison these streams
Don't bury the topsoil
And destroy the good land
For the land and each other
Are all that we have

The land and each other
Are all that we have

For Love Of Kentucky, Stop Destroying Her

by Bob Sloan

In the years I was a social worker, a certain scenario played out so often it always felt like I was sitting through a bad movie for the fiftieth time.

Some guy would beat the hell out of his wife or girlfriend, and after he got hauled off to the slammer, I was sent in to do a "jailhouse evaluation." Usually, before I spoke to the man we called an "offender," I spent some time looking at photos from the emergency room, studying the bruised and bleeding target of his rampage.

Inevitably, during our jail visits, those beasts commenced an endless bleating about how much they loved their women. The third or fourth or tenth time they put the lady in a hospital, they still "loved" her.

Those animals were beneath contempt. Their hypocrisy was wider than the horizon, deep as the sea, limitless as the sky. They were moral midgets.

So are Governor Ernie Fletcher and Congressman Hal Rogers.

So are most members of both houses of the Kentucky Legislature.

So are people running a dozen or more federal and state regulatory agencies.

All of them will seize on any opportunity to talk about how much they "love" Kentucky.

All of them are aiding and abetting mountaintop removal in the eastern coalfields of the Commonwealth. Mountaintop removal is the industrial equivalent of wife-beating.

The Kentucky authors' mountaintop removal tour was two days of travel and observation conceived by Wendell Berry, the dean of Kentucky writers, led by representatives of an organization called Kentuckians For The Commonwealth.

When it was over, all those writers—whose profession was describing things—were at a loss for words when asked what they had seen and heard. They (and I) had seen an environmental holocaust, and we had heard heartbreaking stories from fellow Kentuckians who live in its shadow.

Flying over one mountaintop-removal site, I asked the young pilot to estimate the size of the ravaged ridge system below. "About a thousand acres," he said.

An acre is a football field, more or less.

Below us lay a thousand football fields where not one green thing was left. Not one tree, or bush, or blade of grass. An entire ridge system had been turned the color of gray death.

You can't fly over Death Valley and find a thousand acres where nothing grows.

As writer Ed McClanahan pointed out, you can't find a thousand acres where nothing grows at Chernobyl, site of the world's worst nuclear accident.

Think of it: in parts of Pike County, tap water is poisonous.

We toured the Leslie County farm of Daymon Morgan, whose memories stretch back 79 years. Daymon's healthy hills are surrounded by—and threatened by—the destructive horror of mountaintop removal. Daymon's acres are way back in the country; ten years ago you couldn't hear traffic noise from his porch. Now the diesel roar of trucks and draglines fills the air, with the irritating "beep-beep-beep" of monster machines shifting into reverse.

Underground blasts shake the earth from time to time.

Creeks have disappeared, buried under tons of "valley fill."

That evening at the Hindman Settlement School, we listened as people who live every day amid the destruction of maternal Kentucky hills talked about their lives.

I imagine every one of us took away our own main memory of those stories.

Mine is the young mother who described how she has to make sure her 21-month-old daughter's bath water doesn't get in the baby's mouth. That young mom lives in Pike County. Think of it: in parts of Pike County, tap water is poisonous.

Bill Caylor is president of the Kentucky Coal Association, and what he says about mountaintop removal is propaganda of the crudest sort. Caylor claims the coal industry is "creating land for sustainable development for future generations." And he argues they "reclaim the land by growing grass on it and sometimes adding forms of wildlife."

If he really believes those things, Bill Caylor's the most gullible turkey in the flock. I walked across one of the fields Caylor says the coal industry has "reclaimed." From afar, it appears green, fertile and lovely. Step closer and look down and you see "grass" unlike anything I've ever seen in Kentucky. It's a weird stalk six or eight inches long, growing two or three inches apart, not what you or I would call "grass" at all. The appearance of the field from a distance is a sham and a lie.

So are claims by Ernie Fletcher and Hal Rogers and dozens of politicians that they "love Kentucky." If you love a woman you don't beat her. If you love a place, you don't allow it to be poisoned and polluted, gouged and leveled.

Once, at Midway College, author Kurt Vonnegut said not opposing industrial rape and rapine is morally equivalent to watching Nazis take over Germany, and doing nothing to stop them. When I heard those words ten years ago, they seemed extreme, but the authors' tour convinced me Vonnegut was absolutely right.

If you love Kentucky you'll get involved in stopping mountaintop removal.

Overburden
—with lines from Milosz

by Christina E. Lovin

We learn slowly, we humans:
overburdened with lessons
taught, forgotten, taught
again in the forgetting
until the lesson becomes mere memory
of diminishing returns:

it was the forest that was holding things
together, not the rock and soil
like we once thought,
just as skin with its many layers
bears the burden of the body
with its many layers and without
which the flailed flesh weeps and bleeds,
sinew fails, bones part and lean
to aspects of prayer, part
and fall to groveling, then dust.

Try putting the undone
body, once it is felled
and split, back together:
shove the outside in
and try to give it life as if
those disparate parts belong
together.

There is a saying or there should be:
*Treat the Earth as you would your own
body*, for it is your own body:
nowhere less necessary, nowhere
less precious than the rest:

Tree. Stream. Stone. Steep. Least weed.

This is my body, broken for you:

Tree. Stream. Stone. Steep. Least weed:
*Spring Beauty. Fairy Bells.
Squaw Root. Pennywort.
Vetch. Thistle.*

False Rue.

The migratory water thrush circles over
a plateau of waste gray as ash, as if,
by this, she could find her way home.

There is a saying, or there should be:
*. . . pour millet on graves or poppy seeds
to feed the dead who come disguised as birds.*

Dirty Money

by Kristin Johannsen

"I LOVE COAL."

Drive around the mountains long enough, and you're bound to run across a bumper sticker bearing this sentiment, or something similar. It might be decorating a mud-spattered pickup truck or a late-model Mercedes, but the message is clear.

Or is it? Is the driver really in love with a dirty lump of prehistoric vegetable matter? Does his heart beat faster at the thought of combusting fossil fuel? Is he proclaiming his affection for an arrangement of carbon atoms?

More likely, what the driver really loves is not coal, but what it does for him or her. Coal, we are often told, is essential to Kentucky's economy. Interfere with the coal industry, and you eliminate jobs, destroy communities, and take food out of little kids' mouths. If the future of coal mining is in mountaintop removal, then restrictions on this process will doom countless Eastern Kentuckians to unemployment and a dismal standard of living.

That's what the coal industry would very much like us to believe. But a firsthand look shows a very different picture.

If you ever have a chance to fly over a mountaintop-removal operation, the first thing that will strike you is its staggering scale. You'll see draglines and power shovels bigger than houses scrape away the coal and load it into massive dump trucks. But all of these machines are swallowed up in the enormity of the site, and as you pass over, you'll spot only a few people at work on the field. A handful of human beings are enough to pulverize an entire mountain and shove it aside.

Countless Kentuckians have family ties to coal, with grandfathers and uncles and cousins and parents who worked in mining. For many people's forebears, a steady job in the mines meant a leg up in the world, advantages for their children and grandchildren that they themselves never had.

But these personal ties are becoming ever weaker and more distant. Over the decades, increased efficiency in mining methods means that more and more coal is produced by fewer and fewer people. In 1979, there were 35,902 mining jobs in Eastern Kentucky. By 2003, there were only 13,036. For every three people who once worked the mines, two are now doing something else.

Despite this tremendous drop in coal employment, Kentucky is still turning out nearly as much coal as it ever did—production has gone down only twelve percent in the last two decades. Coal companies favor mountaintop removal because it is the most efficient way to get coal out of the ground—and it cuts their labor costs significantly. In 2003, the average Eastern Kentucky mine worker produced 3.77 short tons of coal per hour in a surface mine—compared with 3.04 tons per hour in an underground mine. Put in human terms, it takes 24 percent fewer workers to produce the same amount of coal by surface mining.

Every form of coal mining, from underground mining to conventional surface mining to mountaintop removal, has seen tremendous increases in productivity in recent years—which means declines in the number of workers needed. Overall, in the Appalachian region, coal mining productivity went up an average of 4.9 percent every year between 1988 and 1997, and it remained 52 percent higher in 2003 compared to 1988.

. . . you would think that Eastern Kentucky communities with coal would be doing much better than communities that lack it.

You would be dead wrong.

But all this is neither here nor there, if you live in Eastern Kentucky and you need a job. Nevertheless, if coal mining really is the economic keystone that the mining industry would like us to believe it is, then you would think that Eastern Kentucky communities with coal would be doing much better than communities that lack it.

You would be dead wrong.

In Eastern Kentucky, there are thirteen counties that produce large amounts of coal—over 500,000 tons per year. Call them the "coal counties"—Bell, Breathitt, Floyd, Harlan, Johnson, Knott, Knox, Lawrence, Leslie, Letcher, Martin, Perry, and Pike. Another twelve eastern counties produce little or no coal—Clay, Elliott, Estill, Jackson, Lee, Magoffin, McCreary, Menifee, Morgan, Owsley, Powell, and Wolfe. We'll call them the "non-coal counties."

So, how do they stack up, economically? Consider the numbers on median family income. (If you ranked all the families in the county from top to bottom by their income, exactly half would earn more than this amount, and half would earn less.) In the thirteen eastern coal counties, median family income is $24,985; in the twelve eastern non-coal counties, it's $24,374. That's not much of a difference, especially when you consider that the median family income for all Kentuckians is $40,939. Coal has little impact on the pattern of family incomes.

Some people will say that this figure includes counties where coal mining isn't that important, so let's look at a place where Coal really is King. In Pike

County, the state's biggest coal producer, 51.8 percent of families must get by on less than $25,000 a year. In the whole of Kentucky, only 37.5 percent of families have such low incomes. In Powell County, which doesn't produce a single lump of coal, the figure was 49.1 percent.

On the whole, information on people's situation in coal and non-coal counties in the 2000 U.S. Census shows very little difference between them. Some numbers are very slightly worse in the coal counties—for example, 38 percent of working-age people have disabilities there, compared with 36.8 percent in the non-coal counties. Other things are very slightly better. In the coal counties, 59.1 per cent of adults have a high school diploma, while 53.3.percent in the non-coal counties have one. But on the whole, the profiles are very similar. In coal counties, 77.9 percent of families are homeowners; in non-coal counties, it's 77.8 percent.

The real shocker is seeing who truly is doing better. Consider a third group of counties in Eastern Kentucky, the ones that are crossed by Interstate Highways 64 and 75. Call these the "interstate counties"—Bath, Boyd, Carter, Fleming, Laurel, Rockcastle, Rowan, and Whitley.

By every economic and social yardstick, they are better off than the coal counties. Median family income in the interstate counties is $32,658—meaning that the "middle" family there earns $8,000 more than in the coal counties. Less than half of families must get by on less than $25,000. Far more people have high school diplomas, far fewer working-age people and children have disabilities, fewer families with kids live under the poverty line—and more people even have telephones.

One important reason for the difference is that major highways help make possible a more balanced economy. Factories and retailers locate there because they can more easily reach a broader market. Workers have access to a wider variety of jobs. Tourists can easily come to explore places like Cumberland Falls, Renfro Valley, and Daniel Boone National Forest. And money flows into the area as capital and stays there, instead of being drained away as natural resources are used up.

In contrast to the coal counties, where communities boom when the mines are hiring and die when they're laying off, areas with more balanced economies are better able to survive economic downturns. When one industry isn't doing well, another will be hiring. And in a balanced economy, more money is circulated through the community by locally-owned businesses, rather than being shipped out in the profits of a faraway coal company.

For 130 years now, we've been told over and over again that coal is good for Kentucky, but the numbers tell a different story. More than 7.8 billion tons of coal have been mined here in that time. But despite the extraction of vast mineral wealth from our land, Kentucky continues its uphill struggle to provide a decent living, a good education, and a clean environment for its people. And the counties of the eastern coalfields, with the richest natural resources of all, remain among the poorest in the entire state.

Today, too many people are still asking, "What would we do if we didn't have coal?" Our question, instead, should be: "What *could* we do if we didn't have coal?"

☒

Results of Burying Headwater Streams

Downstream Damage

> ► Toxic stream chemistry: poisoning of creatures living in or drinking the water
> ► Flooding and erosion
> ► High water-treatment costs
> ► Lost recreational appeal
> ► Lost tourist appeal

Upstream Damage

> ► Increased flooding
> ► Loss of recreational acreage
> ► Loss of animal habitat
> ► Loss of nutrition for fish

—85 aquatic scientists

Letter to the Editor

by Ed McClanahan

Whenever I hear the argument that mountaintop removal creates jobs, I can't help remembering my first encounter, years ago, with a disreputable Henry County neighbor I'll call Pee-Rooney Jimson, who had just pulled two years in the state penitentiary.

"Well, Pee-Rooney," said another neighbor (who knew Pee-Rooney's history better than I did), "what do you plan to do, now that you're out of the pen?"

"Do?" cried Pee-Rooney, in high dudgeon. "I ain't gonna do nothing! they wouldn't let me do what I wanted to do, so I just ain't gonna do nothing!"

"Well," said my friend, "what did you want to do, Pee-Rooney?"

"Why," said Pee-Rooney, indignantly, "I wanted to sell pot and pills to the high school kids!"

I don't mean to seem cold-hearted about the issue of jobs for the guys who blast mountaintops into oblivion and drive overweight coal trucks and gigantic earth-eating bulldozers, but the fact is that these folks need to reconsider what they're doing—to the land, to their heritage, and to the rest of us.

Mankind has probably done more damage to the Earth in the 20th century than in all of previous human history.
                    —Jacques Cousteau

The Elephant in the Overburden

by Whitney Baker

I rediscovered Appalachia in my early twenties. I had once known it as a young boy; my mother took my older sister, younger brother and me to Pine Mountain State Park nearly every Sunday and often Saturday, too, while my father played golf. I was prepubescent and shy; I was bright and curious, but bent toward melancholy. Being in the woods was not a no-brainer for me, it was what I would call a low-brainer; my chatter and worry and fear would scuttle out of my ears and hide itself beneath the forest rot, and be eaten there. The loud silence of the wood filled the emptiness of my young mind, and I was breathing easier, and hearing myself in a body, a-hum. After those years, and their pleasant weekends, I did what we do, I grew, I got immersed in school, I got immersed in television, I discovered the idea of being "rich", and rotting logs were not involved.

All through high school I wanted out, out, out. I wanted to fit myself into one of those Ralph Lauren scenes, with all the terriers and beautiful, disappointed girls in navy and pink plaid. And I got out. I went to Centre College in Danville, Kentucky, and aimed to figure it, how to be who I wasn't, how to make my hometown seem like a charming accident, how to make it seem my mother had dropped me there like a clean handkerchief on her brief missionary trip from Martha's Vineyard. I failed utterly. Then my father died, and I was back home, sleeping on the couch, broke like we had always been, and really, really angry. My dream of transformation via geographical cure had been shattered. I found myself facing the painful truth that I had dreamed up much of my life, a dream energized by the myths of TV, by rent-free room and board, and by an incessant chatter always begun with the phrase, "When I get out of here..." But my trip to Centre, so bountiful with ostensible opportunity, proved I was fundamentally confused. It became necessary that I return home.

It was an instructor from the community college (not native to the area) that told me we were going camping. We had made friends golfing, of all things, and soon we were spending time on the mountain. He was not, and is not, an ordinary human being; in addition to being very smart, charismatic and incredibly athletic, he was grave. My early guided trips up into that place were not safe, nor well lit, and neither mapped nor mappable. We began

176

encountering the mountain (when I say mountain I mean the knowledge of it; to know it from a distance is exactly as intimate as knowing a church by seeing it from the outside) on bicycles, surprisingly. There is a three or four mile lane-and-a-half, ever-winding asphalt road from the visitor's center at Cumberland Gap National Historical Park down in the Yellow Creek basin at Middlesboro up to the crest of a massive rock outcropping overlooking the Cumberland Gap, and beyond, called the Pinnacle. From the designated overview, one can easily see Tennessee, Virginia and Kentucky. Perhaps, as some still insist, the Smokies are occasionally visible. That road is a daytime jaunt for many a visitor; my friend suggested we wait until the city slept and throw our ten speeds over the locked gate to bike up the mountain, sometime after midnight, next full moon. I thought it absurd on its face. You wouldn't be able to *see*. You'd be breaking the rules. You'd have to ride down. It would be too hard. Most of all, why? But I did it, because he had something I wanted.

His brother, older by a couple of years than my new friend, had been killed in Vietnam. He hardly spoke of it, almost never, but he had told me that they sang Beach Boys harmonies, dove into creeks with their burning cigarettes turned back into their mouths to keep them dry. I could see he had survived that death without losing his soul, and could still love the world, though he was clearly suffering and would always suffer. So I knew I would follow that man into that dark.

In hindsight I can see it was the perfect way to begin to get to know the mountain. I met it under my legs, in my legs, in my exhaustion in my lowest gear, feeling nothing but the blind arc of one rise after another, distilled by the road into a smooth and pure translation of massive distension out and away from flat, safe, ordinary earth. It was dark; the moon only sifted silver down onto the mystery of the dark woods and visible pieces of road. There was no color, only light kissing dark, etchings of a thousand thousand lines, my eyes the only parameter in the infinity of information, the world I could see ending with the end of my ability to focus. Frequently, though, there would be a confession by light of allegiance to a thing, for instance those are clearly trees rising up, that is the edge of the road for a bit, that is a fern. There was a mediator at work, deep in my recognition, keeping me in the world of bodies, of trees or only the pedals pushing back, and it kept me from being overwhelmed, or lost. That first night was very, very long and I was very, very tired, but I kept pedaling. Though I was only a few miles from home I was far out, on the edge of knowing. I was scared in a new way that was beyond fear for my body; I was afraid for the known because there was so little I recognized about me, about the darkness, and about persevering. It took a long, long time. Every few hundred feet I knew it was over, that I was seconds away from giving up, but my friend was somewhere up there, and I didn't want to lose him, and I didn't want to go back down there where I would keep watching TV, waiting for it to be exciting again. My father was dead and we hadn't parted on good terms. I apologized to him with his ventilator down his throat, and his hands strapped to the bed. Much more than that, though, I didn't understand what I was doing in this world, why what I wanted the world to be seemed so different from what it was, I didn't understand why so many people and places around me were broken and hurting. And I would not, would not give up on those questions. So I pedaled with whatever did come, from places I never knew, or had forgotten. And I didn't *make* it. Yes, I arrived at the top, and we rode down, that being like releasing a bird to fly from your hands and out your window, a humble, pleasant surrender. I didn't make it because that is the language of victory over defeat. The top had nothing to do with that. The trip deepened where I gave up to giving up, in the densest grinding pushes against the will of my quitting. When I set against that will with my soul-hunger for that unknown, it became known, then and there, in the affirmation from inside, where it was darkest of all, home at that moment to only one thing, disguised as a word, *yes*, then that sound was joined, in harmony, by forest's audible silence, answering its own questions in the language of leaves in small winds, and something small taking a hop, and in crickets. It did not last. With relief returned my mind, with the mind's joy, the mind's meanderings. But it didn't need to last, for then I knew: I had pushed against the dark chaotic failure of my self out in the simple wild of a natural and ungoverned place, away from what I took to be mine, my daily world, and I remained, and was affirmed.

I began to go to the mountain with a constancy, sometimes with friends, but mostly alone. What happened for me on Cumberland Mountain had to happen in an old and rich and dangerous place, because what was at stake was the world. I was, at times, deeply suicidal until that night. Since then I have known that there is no self separate from the extant world, and that the singular myth of suicide, that it can relieve *me*, cannot be true. The beauty of Appalachia is that it knows so much fecundity in perpetual failure, and dying. Go sit at a place there, and be quiet, and long. Eventually, you will know how many things have fallen there, and broken there. If it takes many hours, or days, or months, or years, it is because of you, because your head is too loud, and your eyes are too full. The place is always ready to be known, it doesn't bother with you in the abstract. It doesn't bother. It isn't an it. Where you walk, that impact happens, it myriad consequences, nothing more. For the place, you only exist in your doing. There is no purpose. No impression or intimidation. Yell. Scream you are cheating on your husband. Open your wallet and show. Nothing. Maybe a scarlet tanager will fly to a different maple. That benign disinterest can only be found in the wild, or outer space, where there is no air. So many things have broken. First, though, they grew, or were formed, and thrust. Forever, ice cracks the house-sized stones to fist-sized stones to sand, and the ice is gone. Roots unfurl the wet fallen bark and lick it for needful things, and are gone. Shit catches rain and is held up again in leaves. Where there is wetness there is that which loves wetness. On the huge lip of stone cast into the sun by the ridge in a wave arced out and above hemlock in grove, huckleberry. Though there is no place in any hollow, and no "where" on any hill our words can circumscribe, everything in its place.

Joseph Campbell says the root of the word religion is the Latin *religio* meaning "linking back," the American Heritage Dictionary has *religio* as following *religare* meaning "to tie fast." Meanwhile, now, and for many years, the dominant tone of a frequent and very powerful world psyche is "to link out," or "unfasten." There is thick amongst us the faith that *there* is, by default, preferable to *here. Here* has become a problem to overcome. We can literally see now that the world is spherical, and finite its resources, but that is so terrifying. *Not one person* I know goes on confident that the earth will just keep giving without end. We know that something has to change. And we are very, very young and clumsy with this knowledge. We are, in fact, so scared that we often just hum to ourselves to make it go away. Some of us are turning to a second coming, others to the second season of Crime Scene Investigation, others to saying something pessimistic in our jealousy of others' seemingly more pleasant diversion. All of us are tending toward the abstract, because there we can try to make the rules. We can try to be flip and ironic and say, "The world is going to end anyway, might as well…" But the abstract mind resents what complicates its fantasy. For example, if one wants to fantasize that entertainment is the goal of recreation and can be perfected by research and marketing and accumulation of data, Disneyworld is better than horseshoes in the backyard, talking about Marxism is better than mowing your elderly neighbor's lawn, Country Music Television is better than Uncle Nate on the banjo. "Country Music Television"— listen to the three-part expansion into the abstract: first, a word, "country", for that which cannot, of course, be abstracted, but is at least referred, then, "music," at times a most beautiful human invention, but an invention nonetheless, and finally, devastatingly, "television": the awful compression into crap. Relating to people and land is messy and complicated. The slow parts have not been edited; nothing has been screened or run by a committee. Coal compresses into diamonds, dazzling to be sure, but worthless; similarly, country becomes television: worse than worthless. This is not an aside but the gut of where we are and aren't as inhabitants of earth. Here is the pathology of our issue: are we being in it, or waiting around to be relieved of it?

This abstracted projection of meaning outward and away is, like the elephant in the living room, the elephant in the overburden. If the body, both human and earthen (indistinguishable, I believe), stinky but also pleasant, whatever you may call it, so blatant and undeniable in Appalachia, is something to be transcended, to be waited out, too boring to sit around and "look at," then yes, strip the mountain or don't, whatever. I love the traditional song popularized by the Carter Family, "This World is not My Home:"

> *This world is not my home, I'm just a-passing through*
> *My treasures and my hopes are all beyond the blue*

But I love it with my daytime brain, my thinker, which finds no peace. The song is a gift to my restless spirit from my soul, which needs no such song, nor a heaven that depends on the passage of time to arrive. The soul never worries about comings and goings, it is at peace and unknowable and somehow both within me and without me always. Many mistake the daytime brain as their truest self, as that to which Christ, or any prophet, spoke. The solution to wanton environmental destruction, destruction without appropriate sadness, without proportion, is a soul reckoning, a religious revolution. And that last sentence is a dangerous sentence, because it is words, it is abstract, and cannot be mistaken for the thing itself, lest it become enemy of the thing itself. Soul reckoning means churches must stop looking and feeling like Greyhound stations waiting for completion of a rail to Outer Space. Our energy consumption must stop being malicious (see the Hummer). Our free time must involve movement and engagement, not sixty years of "freeze tag" on the couch. We are obliged to the utmost decency and humility before what we see, where we are, what we wear when we are naked, and what we remember when we have forgotten all the facts. There is a knowing beyond words. There is a prayer that *is* silence, free of the brain's interference. It is stewardship. We have nearly lost it. The decision to envision and invoke such a life contributes to a healthier planet, but it does not insure one. We cannot act based on whether we think we will "win." We must abandon the victory-based mindset.

A founder of the Holiness Christian movement said, when speaking to leaders of the church upset over the loss of membership, "Just focus on the Holiness and the rest will take care of itself." For those of us, scattered and relatively powerless, facing the mindless, hellish chewing and spitting out of our mountaintops, those of us beneath the rain of stone that follows the blast, our "Holiness" is *what*? If we opponents of the practice of mountaintop removal mining can call ourselves "we," what unites us? It is our ignorance. If we had the big answers, and they were: yes, heaven is coming, you'll be welcome, and it is all to happen soon, or, we'll invent a machine that keeps our brains alive in jars and it will feel great all the time, we could dismiss the ancient land, blissfully cruising until, inevitably, all is turned to wormwood and fire, or what's left of it is uploaded onto a hard-drive, then, if we wish, downloaded into our jars. Its complexity would be no more or less interesting than a forty-year-old phone book, quaint and imitable as the fake tree at Disneyworld. We spend billions of dollars on virtual *reality*. Is reality missing? No, only our ability to see it has diminished. The living earth must be reckoned with.

The natural world is our guide. It knows things we have forgotten, or never knew. The more natural, and more complex, the more it can teach us. Not simply because the cure for cancer might be spliced from the lady slipper's toe, or because we might still salvage some amazing authentic Cherokee pipe. The lessons are in making due, in finding one's right place, in embracing what is broken and what has died, in relating to one's surroundings, in responding appropriately to circumstance. The mountains and forests, like all great and awesome landscapes, forgive us our endless grief, our naïve hope, and the boundless ignorance which is our grace. We can return to the old places to start again. We will never be perfect— that idea is the property of the architects. Inevitably, we will mark the earth. Furthermore, it is right for us to harvest a share of earth's bounty, but it is best to do it in a sustainable manner. Terribly, when we destroy these mountains with mountaintop removal mining, what is gone was priceless and what comes in its stead is wrong. Abstraction and extraction have a place. However, we evolved in imperfect harmony with the natural earth, able, unlike any other being, to think beyond instinct, but unable, perhaps fatally, to reproduce or manufacture what makes us at home in the only place that could have us. At this crucial time, we must listen with our deepest attention for each next right step.

We are working for a day when Kentuckians—and all people—enjoy a better quality of life. When the lives of people and communities matter before profits. When our communities have good jobs that support our families without doing damage to the water, air and land. When companies and the wealthy pay their share of taxes and can't buy elections. When all people have health care, shelter, food, education, clean water and other basic needs. When children are listened to and valued. When discrimination is wiped out of our laws, habits and hearts. And when the voices of ordinary people are heard and respected in our democracy.

—Kentuckians For The Commonwealth

Looking for Hope in Appalachia

by Erik Reece

From the top of a working fire tower in Eastern Kentucky's Robinson Forest, I can look out over two very different landscapes, two very different economies. Looking south, I see 12,000 acres of green shouldering hills. Over sixty species of trees cover these ridge-lines where they, among other things, provide crucial habitat for deep-forest dwellers like the rare cerulean warbler, and they supply these watersheds with the cleanest streams in Kentucky.

Turning north from the fire tower, I see 1,000 more acres, but the trees and the ridgetops are gone. The peaks have been blasted away by mountaintop-removal mining, and what remains are gray and brown plateaus—a desert really—where a monoculture of grasses struggles, and largely fails, to grow in slate and sandstone. This landscape is pocked with black silt ponds and gray craters.

These days, it is thought unfashionable, even backward, to talk about laws of nature or to read a philosophy, a morality, into the workings of the natural world. But compare these two economies: the forest's and the strip mine's. The sulfur dioxide that escapes coal-burning plants is responsible for acid rain, smog, respiratory infections, asthma, and lung disease. In 2000, the Clean Air Task Force, commissioned by the Enviromental Protection Agency (EPA), determined that coal-fired power plants account for 30,000 deaths per year. In Kentucky, the number of children treated for asthma has risen almost fifty percent since 2000. Due to acid rain and acid mine runoff, there is so much mercury in Kentucky streams that any pregnant woman who eats fish from them risks causing serious, lifelong harm to the fetus she carries. Furthermore, as we all know but choose to ignore, fossil fuels—coal and oil—are responsible for the carbon dioxide that is making the planet hotter and its weather more volatile. This year, climatologists found record-high levels of carbon dioxide in the atmosphere.

A forest, by contrast, can store twenty times more carbon than cropland or pastures. Its economy is a closed loop that transforms waste into food. In that attribute alone, it is superior to our human economy, where the end of the line is not nutrients, but rather toxic industrial waste. A forest's leaf litter slows erosion and adds organic matter to the soil. Its dense vegetation stops flooding. Its headwater streams purify creeks below. A contiguous forest ensures species habitat and diversity. A forest, in short, does all of the things that the mining and burning of coal cannot—that is its intelligence. Is there design behind this natural intelligence? I have no idea. But I will venture this: the forest knows what it's doing.

Why is our industrial/consumer economy so out-of-sync with the natural economy, and what might be the consequences of this ecological blindness? In his essay, "The Last Americans," Jared Diamond argues that while wealth and conspicuous consumption are certainly signs of social status, they may not indicate success for a society as a whole. In fact, if Mayan civilization (among many others) is an indicator, material prosperity, overpopulation, resource consumption, and waste production are actually signs of a society's impending collapse. Like the coal operators of Appalachia, the Mayans stripped their forests and polluted their streams with silt and acids. Their population spiked steeply in the 5th century. More and more people started fighting over fewer resources. By the time Cortez marched through the Yucatan, a culture that once numbered in the millions was gone. "Why," Diamond asks, "did the kings and nobles not recognize and solve these problems? A major reason was that their attention was evidently focused on the short-term concerns of enriching themselves, waging wars, erecting monuments, competing with one another, and extracting enough food from the peasants to support all those activities." It sounds to me like an unnerving assessment of this century's first five years.

> . . . we should consider what the Nobel Committee decided . . . The Norwegians awarded the 2004 Nobel Peace Prize to a Kenyan woman, Wangari Maathai— for planting trees.

Three days after George W. Bush's re-election, Vladimir Putin signed the global, carbon-dioxide-reducing Kyoto Protocol, thereby ratifying the treaty that could only go into effect if 55 percent of countries producing greenhouse gases agreed to its mandates. And while the United States produces twice as much carbon dioxide as Russia, Bush made it clear in the Presidential debates that he would not sign the Kyoto Protocol because it could "cost American jobs." In other words, short-term decision-making will continue to rule the day though the long-term effects of those decisions could be disastrous.

The re-election of Bush, the rising price of coal, and the quickening pace of mountaintop removal throughout central Appalachia can cause one to despair. Therefore, I think we should consider what the Nobel Committee decided around the same time Americans were deciding in favor of George W. Bush. The Norwegians awarded the 2004 Nobel Peace Prize to a Kenyan woman, Wangari Maathai—for planting trees.

Twenty years ago, Maathai's countryside, like the mountains of central Appalachia today, was on the verge of desertification. Poor rural women had to walk farther and farther to collect less and less wood for heating and cooking. So Maathai mobilized a legion of those women to start planting woodlots, known as green belts, throughout their villages. Soon, there was more shade, less erosion, cleaner water, cleaner air, ample firewood for cooking, and jobs. Maathai's Green Belt Movement started solving problems of poverty, malnutrition, pollution, and women's lack of rights. Maathai brilliantly used the example of a forest community to reestablish human communities across Kenya.

The Nobel Committee's recognition of Maathai's work and the way it has improved the lives of poor communities is an important signal that Appalachia should take very seriously. There is so much abandoned, unreclaimed mine land in the United States that the Federal Office of Surface Mining doesn't even try to account for it in terms of acreage. And one visionary organization, the Appalachian Regional Reforestation Initiative, now has developed a successfully tested plan to bring forests back to strip mines. Based on experiments conducted by Clark Ashby in the 1960s, foresters have found that the best way to reintroduce trees on a strip-mined mountain is simply to leave the soil uncompacted by bulldozers. If haul trucks dumped four-foot-deep rows of spoil across a mine site and planted tree seedlings in those mounds, even without any topsoil, the rate of growth would be double that of seedlings in their native forests. Why? The loose soil gives tree roots plenty of room to sink their tentacles, so the trees grow faster. Within five to eight years, their crowns would start to close, according to Office of Surface Mining Field Manager Patrick Angel.

There is so much abandoned, unreclaimed mine land in the United States that the Federal Office of Surface Mining doesn't even try to account for it in terms of acreage.

This man-made forest would mature faster than a natural forest because it would cut out about fifty years of early successional growth. That is to say, one can start planting late-succession trees like oaks, walnuts and other valuable hardwoods right away. Those can be accompanied by faster-growing pines, which could be harvested sooner for moldings or pulp wood. In a short amount of time, a sustainable, local industry could literally take root on these barren plateaus. And if locally-owned secondary industries like furniture- and cabinet-making took hold, the region's economy might begin freeing itself from a century of colonization by absentee coal barons. Trees, as I've mentioned, sequester much more carbon dioxide from the air than grasses. And pieces of furniture constitute what is known as a "carbon sink"—that is to say, unlike paper or firewood, the carbon in a piece of furniture is going to stay there. As a final recommendation for this form of reclamation, the loose soil piled on the mine sites would collect much more rainwater than compacted valley fills, thus easing problems with flooding.

It is important to emphasize that this plan does nothing to stop mountaintop removal; that, I am convinced, will only happen when enough public outrage shames politicians, regulators, and coal operators into doing the right thing. But this plan for real reclamation of already-mined sites at least returns the forest's intelligence to a landscape that has for too long been wasted by human arrogance.

Who is going to pay for all of this? There is currently $1.7 billion in the federal Abandoned Mine Land (AML) fund—about the same amount of money that was spent to launch the last space shuttle of a mission of little scientific value. Yet virtually none of it is released for reclaiming abandoned mine sites because Congress would rather use the interest from it to pay down our ballooning national deficit. In addition, the coal industry doesn't want the job competition that real reclamation would create. This withholding of funds is unconscionable, especially when you realize that forty percent of Americans' tax dollars are now spent on the nation's war machine. It is time that fiscal values come in line with our often-touted moral values. Not only should the AML money be released to the coalfields, but, given that the price of coal has doubled in the last year, Kentucky's 4.5 percent severance tax (which hasn't risen since 1976) should reflect that increase and be raised at least to Wyoming's level of seven percent. Finally, in Kentucky, the current severance tax is put into a general fund, and over half of it is spent outside the coal counties. However, since it is Eastern Kentuckians who must live with the flooding, contamination, and destruction of roads and homes caused by mountaintop removal, it is only fair that Eastern Kentucky receive all of the money from the severance tax. These changes in reclamation and funding would, I believe, trigger a "New Deal for Appalachia," whereby a generation of local men and women could be trained to reforest and manage abandoned strip mines, and finally address the region's chronic problems of poverty, unemployment, and drug-addiction.

Still, a legion of public-spirited tree-planters will not alter a mindset that favors individual acquisitiveness over community responsibility. In the end, Appalachia is a region that suffers, like the nation as a whole, from a larger spiritual crisis. In our consumer-driven culture, we have chosen no longer to think of the world as God-given. It's too inconvenient. Instead, here in Kentucky, our forests and streams are supposed to be protected by a Department of Natural Resources, because that's all we see them as—simply resources. We seldom see value in the natural world, whether aesthetic or intrinsic; we only see something we can use, even if that means using it up. We no longer see ourselves as part of a greater whole, a world so vast and mysterious that it deserves our reverence alongside our scientific probing. In America today, the "environment" is almost wholly other: We are over here, and it is over there. We are in the air-conditioned mall; it is hot and crawling with bugs. And anyone who prefers the out-there is an "environmentalist," that oddly dressed guy who thinks this diminutive planet might be worth saving.

The 20th century theologian Martin Buber made a crucial and influential distinction between two kinds of relationships: one based on an "I–It" principle, the other on an "I–You" way of thinking. The I–It mindset is reflected in the situation I have been describing: there is a subject and an object and each is alien to, separate from, the other. In the I–You experience, something stranger happens. One loses a sense of the well-defined ego, the self; as a result, the other, the You, begins to seem not so alien anymore. Indeed, in some mysterious way, the You starts to feel like more of the I, and both the I and the You seem caught up in some larger interconnectedness. The 13th century Buddhist master Dogen said, "To study Buddhism is to study the self. To study the self is to forget the self. To forget the self is to be enlightened by all things." Mystics call this experience satori, or enlightenment; modern scientists simply call it ecology. In a forest, trees are photosynthesizing carbon into oxygen every spring. Whether I feel a spiritual connection to that forest or not, it is keeping me alive. It is part of me whether I want to admit it or not.

The land is one organism—American naturalist Aldo Leopold thought this to be "the outstanding discovery of the 20th century." James Lovelock and Lynn Margulis reaffirmed this discovery in the early 1970s, when their research into the temperature, composition, and oxidation rate of the atmosphere found that the Earth does indeed act like one self-regulating macro-organism, wherein all of the parts are working interdependently for the health of that larger organism. They even gave it a name, Gaia—Greek for "Mother Earth." For Leopold, whose influence on wildlife wilderness protection in the US can hardly be overestimated, the health of that organism depends on two things: stability and diversity. The stability of an integrated forest leads to the accumulation of soil fertility; soil fertility leads to biological diversity; biological diversity leads to "one humming community of co-operators and competitions, one biota [plant and animal life in one area]." Of any large community of flora and fauna, Leopold said we must ask two questions: (1) Does it maintain fertility? (2) Does it maintain a diversity? These two questions form the basis of what Leopold called a "land ethic," which if successful, would "change the role of Homo sapiens from conqueror of the land-community to plain member and citizen of it."

No one who felt a responsibility to other citizens within a community would destroy its water, homes, wildlife, and woodlands. The difference between conquerors and community is the difference between the words "economy "and "ecology." Both come from the same Greek root, "oikos," meaning "the study of homes." But only ecology has remained such a study. A true case of home economics would, as Leopold said, make sure that the place called home maintains its health and stability. To create an environment where mudslides, flooding, black water and slurry spills are common undermines a community's health. To bulldoze and burn a renewable resource, trees, and replace them with huge valley fills will not ensure stability. To tear a nonrenewable resource from the ground to provide a short-term economic gain for the few and long-term environmental destruction for the many is undemocratic, unsustainable, and stupid.

We are, unfortunately, a nation that values technology and wealth much more than we value community. The 20th century was a Faustian gamble that combined industrialism and greed to make us cash-rich and resource-poor. As E.O. Wilson wrote, "We and the rest of life cannot afford another hundred years like that." If our species is to make it through this century, the forces of science and technology must be tempered by two other forces—ethics and aesthetics. As Leopold observed, all philosophies of ethics, from Aristotle on down, are actually based on this ecological principle: "that the individual is a member of a community of interdependent parts." And as the cave art at Lascaux makes brilliantly clear, we are a species that has evolved to find beauty in the natural world. This trait serves—or should serve—an evolutionary purpose: we love what we find beautiful, and we do not destroy that which we love. What a strip job demonstrates, then, is the absence of any ethic or aesthetic. It is more than a moral failure; it is a failure of the imagination, a failure to understand the employment and energy alternatives that would preserve the integrity and the beauty of Appalachia.

When Israel came out of Egypt . . . the mountains
skipped like rams; and the little hills like lambs.
—Psalm 114

Let There Be a Home Always

by Charlie Hughes

Let the road stretch unending before me.
Let the trees become a green blur.
Let the speedometer needle dip below the dash
and the temperature gauge rest forever on cool.
Let my destinations rise before me
like mountains from the mist.
Let each arrival be a celebration.

Let the road cling to the mountainside.
Let the trees touch their arms above me.
Let the old Chevy enjoy each switchback
and breathe easy on the downgrade.
Let each village and burg come upon me sudden
after rounding a bend.
Let each arrival be a celebration.

Let the road bend snaky along the creekbank.
Let trees cast shadows on the yellow water.
Let heat shimmer on the asphalt
and coal dust lift from the kudzu.
Let each roadside general store and gaspump
promise Moonpies and RCs.
Let each arrival be a celebration.

Let the road disappear into the night.
Let the trees embrace my headlights.
Let wind enter each window—
evaporate the sweat from each brow.
Let there always be a home
where we arrive at midnight.
Let each arrival be a celebration.

Don't Just Stand There, DO SOMETHING!

by Jerry Hardt

In November 1988, 82.5 percent of Kentuckians approved a constitutional amendment to end the worst abuses of the broad form mineral deed.

While coordinating that campaign, Kentuckians For The Commonwealth found that it wasn't so necessary to convince voters which way to vote as it was merely to let people know what was going on. Broad form deeds affected residents in only about a dozen counties, and strip mining occurs in only about a third of Kentucky's 120 counties. Yet the coal companies' ability to strip-mine and destroy people's land without their permission and against their wishes was such a blatant abuse of power that Kentuckians everywhere were eager to put an end to this injustice, and they did so convincingly. The ballot initiative won in every county in Kentucky.

Mountaintop removal and valley fills are no less of an injustice and abuse of power. And the challenge is similar—to let Kentuckians know what is happening to the people of the coalfields and to the state we all love. And then to take action. One critical difference between our present battle and that of 1988 is that this issue really does affect every person who lives in Kentucky.

The broad form deed victory was made possible by what had happened earlier in 1988, as well as what had happened in 1984, and in 1974, and in 1965—incidents of resistance and action described in an earlier essay in this book, "The Long Struggle." Each of these public actions to make the coal industry more accountable built upon those that had taken place earlier—or more accurately, upon the organizing, educating, lobbying, networking, and building of momentum that took place behind the scenes and over many years. That's what we're about right now: building—or perhaps rebuilding—the movement that will put an end to mountaintop removal and valley fills. And again, it will take Kentuckians everywhere to bring about this victory.

We also realize that an even broader movement is needed to bring about an economic system that is not dependent upon the destruction of the land and the oppression of the people. More than 100 years of abuse at the hands of the coal industry have proven that ending various of the industry's most outrageous abuses, while improving the lives of those affected, does little to change the structures of oppression and the persistence of poverty that are the

190

hallmarks of the coal industry's presence in a community. We must build a future beyond coal.

But let's take one step at a time. What needs to happen next and what can each of us do? We must takes steps immediately to stop mountaintop removal and valley fills. These practices are unnecessary for meeting the state's and the nation's energy demands—even our specific demand for coal. But these practices also reflect a choice that has been made, placing profits ahead of people. That is a choice that needs to be unmade.

It is unlikely that this issue will ever come before the state's voters as a direct question, as the broad form deed did. Nonetheless, the outcome of this issue will be determined by the way we all vote. The governor and legislators have the power to put some reasonable controls on the mining of coal and then to enforce those controls. In fact, just getting a governor and legislators who are not beholden to King Coal would be a major step. We must help make this happen wherever we live and vote.

Legislators who do the bidding of the coal industry at the expense of the health of our people, land, and economy need to pay the price on Election Day.

Additionally, we can support legislation to stop these practices. A bill was introduced in the 2005 General Assembly to stop valley fills, and it will be introduced again in 2006 and in future legislative sessions until it passes. The bill did not pass in 2005 largely because the chair of the House committee to which it was assigned is in the pockets of the coal industry. And the majority floor leader, who controls the flow of legislation, is a coal industry employee.

These men will not let mountaintop-removal and valley-fill legislation move through the House unless and until legislators from all over the state demand that they do so—and stake some of their political capital on making it happen. As concerned citizens, we are the ones who will motivate our legislators to make this action a priority.

A troubling incident in the 2004 legislature shows how far we still have to go. In a House committee hearing on a bill that would promote an incentive for the development of renewable energy, the Fletcher administration helped weaken the bill before working for its passage. The coal industry testified against it. Several legislators who voted for the bill still felt it necessary to fumble through apologies to the coal industry for supporting any legislation that the industry did not like.

If Kentucky is to move beyond our unhealthy dependence on coal, our citizens must make it clear that they will not tolerate such incidents and attitudes. Legislators who do the bidding of the coal industry at the expense of the health of our people, land, and economy need to pay the price on Election Day.

What had happened earlier in 1988, before the November vote, was that the General Assembly agreed to put the issue of the broad form deed on the ballot. In the course of their deliberations about whether or not to do so, every legislator heard from his or her own constituents, as well as voters from all around the state. People from every corner of the state voiced their urgent demand for the ballot initiative to go forward. Such public pressure needs to make itself felt again. Every legislator needs to hear from constituents demanding that mountaintop removal and valley fills end in Kentucky.

Then we need to support initiatives—public, private, and legislative—to build a sustainable and survivable energy supply that moves us beyond our unhealthy dependence on coal and the other fossil fuels. In this transition, we all must be willing to provide a focused effort toward economic development and diversification for those coal counties that need it most, which happen also to be the ones that historically have been the most overlooked. We need a total transformation of the economies of counties where coal is most heavily mined. So what must we do?

▶ We must vote, and vote wisely. The way coal companies are allowed to treat the land and people must become a campaign issue. Candidates must hear our concerns and know we expect substantive answers.

▶ We must communicate with elected officials and hold them accountable for their actions. That means we must also stay informed ourselves.

▶ We must support those individuals and groups who are out front organizing and those working behind the scenes to build the movement. These are the people who will keep us informed, help create options, and let us know when action is most needed. The participation of every concerned Kentuckian is essential.

▶ All people, particularly people of conscience (for this is a moral issue as well as an economic one), must speak out. Allowing mountaintop removal is a choice—one that we all participate in, knowingly and unknowingly, until we say NO. Let's say it.

▶ We must become conscious of the ways we use energy. It is our relentless demand for more and cheaper energy that is used as justification for stripping coal in the quickest and cheapest way possible. We must remove this justification.

▶ We must demand a green-energy option from our local utilities. If the demand is there, they will answer it.

▶ We must ask our electricity suppliers if any of the coal they use is mined using mountaintop removal. We must ask them to stop using coal mined by this method. Just as a Tennessee Valley Authority contract decades ago spurred the demand for strip-mined coal, wouldn't it be great if utilities' demand for non-stripped coal helped put an end to the practice?

► We must demand state and national energy policies and programs that emphasize renewable energy sources and conservation. Government must make available the resources to move us toward a green-energy future.

► We must have a vision. We must believe that what is unacceptable, as described in this book, does not have to continue. A vision will help us see the path to change, and that will give us the hope and courage to move forward.

Hymn

by Tony Crunk

Going down
the valley

down
the valley

only a spark
for warmth

among the shadows.
Going to lie down

in darkness
mother

here
and here after.

Going to rise
again

rise again
as stars

begin to fall
like bread from heaven.

Going to take up
my body

children
walk this burning earth

walk it
burning.

Afterword

by Wendell Berry

At the end of a book attempting to deal with an enormity so staggering as the human destruction of the Earth, it is difficult to resist the temptation to write out a "vision of the future" that would offer something better. Even so, I intend to resist. I resist, not only because such visions run a large risk of error, but also out of courtesy. A person of my age who dabbles in visions of the future is necessarily dabbling in a future that belongs mostly to other people.

What I would like to do, instead, if I can, is help to correct the vision we Kentuckians have of ourselves in the present. In our present vision of ourselves, we seem to be a people with a history that is acceptable, even praiseworthy, a history that we are privileged to inherit uncritically and with little attempt at rectification. But by the measures that are most important to whatever future the state is to have, ours is a history of damage and of loss.

In a little more than two centuries—a little more than three lifetimes such as mine—we have sold cheaply or squandered or given away or merely lost much of the original wealth and health of our land. It is a history too largely told in the statistics of soil erosion, increasing pollution, waste and degradation of forests, desecration of streams, urban sprawl, impoverishment and miseducation of people, misuse of money, and, finally, the entire and permanent destruction of whole landscapes.

Eastern Kentucky, in its natural endowments of timber and minerals, is the wealthiest region of our state, and it has now experienced more than a century of intense corporate "free enterprise," with the result that it is more impoverished and has suffered more ecological damage than any other region. The worst inflicter of poverty and ecological damage has been the coal industry, which has taken from the region a wealth probably incalculable, and has imposed the highest and most burdening "costs of production" upon the land and the people. Many of these costs are, in the nature of things, not repayable. Some were paid by people now dead and beyond the reach of compensation. Some are scars on the land that will not be healed in any length of time imaginable by humans.

The only limits so far honored by this industry have been technological. What its machines have enabled it to do, it has done. And now, for the sake

197

of the coal under them, it is destroying whole mountains with their forests, water courses, and human homeplaces. The resulting rubble of soils and blasted rocks is then shoved indiscriminately into the valleys. This is a history by any measure deplorable, and a commentary sufficiently devastating upon the intelligence of our politics and our system of education. That Kentuckians and their politicians have shut their eyes to this history as it was being made is an indelible disgrace. That they now permit this history to be justified by its increase of the acreage of "flat land" in the mountains signifies an indifference virtually suicidal.

So ingrained is our state's submissiveness to its exploiters that I recently heard one of our prominent politicians defend the destructive practices of the coal companies on the ground that we need the coal to "tide us over" to better sources of energy. He thus was offering the people and the region, which he represented and was entrusted to protect, as a sacrifice to what I assume he was thinking of as "the greater good" of the United States. But this idea, which he apparently believed to be new, was exactly our century-old policy for the mountain coalfields: the land and the people would be sacrificed for the greater good of the United States—and, only incidentally, of course, for the greater good of the coal corporations.

The response that is called for, it seems to me, is not a vision of "a better future," which would be easy and probably useless, but instead an increase of consciousness and critical judgment in the present. That would be harder, but it would be right. We know too well what to expect of people who do not see what is happening or who lack the means of judging what they see. What we may expect from them is what we will see if we look: devastation of the land and impoverishment of the people. And so let us ask: What might we expect of people who have consciousness and critical judgment, which is to say real presence of mind?

We might expect, first of all, that such people would take good care of what they have. They would know that the most precious things they have are the things they have been given: air, water, land, fertile soil, the plants and animals, one another—in short, the means of life, health, and joy. They would realize the value of those gifts. They would know better than to squander or destroy them for any monetary profit, however great.

Coal is undoubtedly something of value. And it is, at present, something we need—though we must hope we will not always need it, for we will not always have it. But coal, like the other fossil fuels, is a peculiar commodity. It is valuable to us only if we burn it. Once burned, it is no longer a commodity but only a problem, a source of energy that has become a source of pollution. And the source of the coal itself is not renewable. When the coal is gone, it will be gone forever, and the coal economy will be gone with it.

The natural resources of permanent value to the so-called coalfields of Eastern Kentucky are the topsoils and the forests and the streams. These are valuable, not, like coal, on the condition of their destruction, but on the opposite condition: that they should be properly cared for. And so we need, right now, to start thinking better than we ever have before about topsoils and forests and streams. We must think about all three at once, for it is a violation of their nature to think about any one of them alone.

The mixed mesophytic forest of the Cumberland Plateau was a great wonder and a great wealth before it was almost entirely cut down in the first half of the last century. Its regrowth could become a great wonder and a great wealth again; it could become the basis of a great regional economy—but only if it is properly cared for. Knowing that the native forest is the one permanent and abundant economic resource of the region ought to force us to see the need for proper care, and the realization of that need ought to force us to see the difference between a forest ecosystem and a coal mine. Proper care can begin only with the knowledge of that difference. A forest ecosystem, respected and preserved as such, can be used generation after generation without diminishment—or it can be regarded merely as an economic bonanza, cut down, and used up. The difference is a little like that between using a milk cow, and her daughters and granddaughters after her, for a daily supply of milk, renewable every year—or killing her for one year's supply of beef.

A forest ecosystem, respected and preserved as such, can be used generation after generation without diminishment—or it can be regarded merely as an economic bonanza, cut down, and used up.

And there is yet a further difference, one that is even more important, and that is the difference in comprehensibility. A coal mine, like any other industrial-technological system, is a human product, and therefore entirely comprehensible by humans. But a forest ecosystem is a creature, not a product. It is, as part of its definition, a community of living plants and animals whose relationships with one another and with their place and climate are only partly comprehensible by humans, and, in spite of much ongoing research, they are likely to remain so. A forest ecosystem, then, is a human property only within very narrow limits, for it belongs also to the mystery that everywhere surrounds us. It comes from that mystery; we did not make it. And so proper care has to do, inescapably, with a proper humility.

But that only begins our accounting of what we are permitting the coal companies to destroy, for the forest is not a forest in and of itself. It is a forest, it can be a forest, *only* because it comes from, stands upon, shelters, and slowly builds a fertile soil. A fertile soil is not, as some people apparently

suppose, an aggregate of inert materials, but it is a community of living creatures vastly more complex than that of the forest above it. In attempting to talk about the value of fertile soil, we are again dealing immediately with the unknown. Partly, as with the complexity and integrity of a forest ecosystem, this is the unknown of mystery. But partly, also, it is an unknown attributable to human indifference, for "the money and vision expended on probing the secrets of Mars . . . vastly exceed what has been spent exploring the earth beneath our feet." I am quoting from Yvonne Baskin's sorely needed new book, *Under Ground*, which is a survey of the progress so far of "soil science," which is still in its infancy. I can think of no better way to give a sense of what a fertile soil is, what it does, and what it is worth than to continue to quote from Ms. Baskin's book:

> . . . a spade of rich garden soil may harbor more species than the entire Amazon nurtures above ground. . . . the bacteria in an acre of soil can outweigh a cow or two grazing above them.

> Together [the tiny creatures living underground] form the foundation for the earth's food webs, break down organic matter, store and recycle nutrients vital to plant growth, generate soil, renew soil fertility, filter and purify water, degrade and detoxify

200

pollutants, control plant pests and pathogens, yield up our most important antibiotics, and help determine the fate of carbon and greenhouse gases and thus, the state of the earth's atmosphere and climate.

By some estimates, more than 40 percent of the earth's plant-covered lands . . . have been degraded over the past half-century by direct human uses . . .

The process of soil formation is so slow relative to the human lifespan that it seems unrealistic to consider soil a renewable resource. By one estimate it takes 200 to 1,000 years to regenerate an inch of lost topsoil.

And so on any still-intact slope of Eastern Kentucky, we have two intricately living and interdependent natural communities: that of the forest and that of the topsoil beneath the forest. Between them, moreover, the forest and the soil are carrying on a transaction with water that, in its way, also is intricate and wonderful. The two communities, of course, cannot live without rain, but the rain does not fall upon the forest as upon a pavement; it does not just splatter down. Its fall is slowed and gentled by the canopy of the forest, which thus protects the soil. The soil, in turn, acts as a sponge that absorbs the water, stores it, releases it slowly, and in the process filters and purifies it. The streams of the watershed—if the human dwellers downstream meet their responsibilities—thus receive a flow of water that is continuous and clean.

. . . we are not talking here about the preservation of "the American way of life." We are talking about the preservation of life itself.

Thus, and not until now, it is possible to say that the people of the watersheds may themselves be a permanent economic resource, but only and precisely to the extent that they take good care of what they have. If Kentuckians, upstream and down, ever fulfill their responsibilities to the precious things they have been given—the forests, the soils, and the streams—they will do so because they will have accepted a truth that they are going to find hard: the forests, the soils, and the streams are worth far more than the coal for which they are now being destroyed.

Before hearing the inevitable objections to that statement, I would remind the objectors that we are not talking here about the preservation of "the American way of life." We are talking about the preservation of life itself. And in this conversation, people of sense do not put secondary things ahead of primary things. That precious creatures (or resources, if you insist) that are infinitely renewable can be destroyed for the sake of a resource that to be used must be forever destroyed, is not just a freak of short-term accounting and the externalizing of cost— it is an inversion of our sense of what is good. It is madness.

And so I return to my opening theme: it is not a vision of the future that we need. We need consciousness, judgment, presence of mind. If we truly know what we have, we will change what we do.

CONTRIBUTORS

Whitney Baker is from Middlesboro, KY. He moved to Lexington after graduation from Lincoln Memorial University in Harrogate, TN, with degrees in Studio Art and Mathematics. At L.M.U. he was awarded the Ross S. Carter Creative Writing Award and the award for the best graduating art student. Mr. Baker works as a garden designer and consultant, as well as a painter and graphic artist. Formerly the poetry editor of *Wind Magazine*, he occasionally reads from his own work in and around Lexington.

Artie Ann Bates is a child psychiatrist who works at a child advocacy center in Hazard, Kentucky. She is a native of Blackey, Kentucky, in Letcher County. Her children's book, *Ragsale*, was published by Houghton Mifflin in 1996, and she has essays in the women writers' anthologies *Bloodroot* and *Listen Here*.

Wendell Berry, poet, novelist, essayist, and farmer, is the author of such admired works as *Jayber Crow, The Unsettling of America, The Memory of Old Jack*, and *A Place on Earth*. Winner of a Guggenheim Fellowship and a Lannan Foundation award for nonfiction, he lives and farms in Henry County with his wife, Tanya, near their children and grandchildren.

George Brosi was raised in Oak Ridge, Tennessee. He has been a bookseller specializing in Appalachian titles for more than a quarter-century. He is also the editor of *Appalachian Heritage*, a region-wide literary quarterly published by Berea College. A book he recently co-edited, *No Lonesome Road: the Prose and Poetry of Don West*, received an American Book Award for 2004.

Warren Brunner, a Berea photographer, has been capturing evocative images of Appalachia since the 1960s. In 1994, he collaborated with Loyal Jones on a collection of illustrated essays, *Appalachian Values*.

Geoff Oliver Bugbee grew up in Louisville. For the last 10 years, he has worked as a newspaper and magazine stringer in the U.S. while devoting himself to documenting international social issues. His work has appeared in many of our nation's most prominent newspapers and magazines.

Steven R. Cope is an author and musician originally from Menifee County, Kentucky. His works include four volumes of poetry, *In Killdeer's Field, Clover's Log, CROW!—The Children's Poems*, and *The Furrbawl Poems*, the novel *Sassafras*, and the fable/story collection *The Book of Saws*. He has also edited several noted and important works.

Tony Crunk is a native of Hopkinsville, Kentucky. His first collection of poetry, *Living in the Resurrection*, was the 1994 selection in the Yale Series of Younger Poets. His work has appeared in such journals as *The Paris Review, The Georgia*

Review, The Virginia Quarterly Review, and *Poetry Northwest.* He has also published two books for children, *Big Mama* (1999) and *Grandpa's Overalls* (2000).

Charles Bracelen Flood is the author of eleven books of fiction and non-fiction, including *Lee: the Last Year; Hitler: the Path to Power,* and *Rise, and Fight Again: Perilous Times along the Road to Independence,* which won an American Revolution Round Table Award. He has lived for the last thirty years in Richmond, Kentucky. His *Grant and Sherman: the Friendship that Won the Civil War* is being published this autumn by Farrar, Straus & Giroux.

Lucy Flood grew up in Madison County, Kentucky. She is a recent graduate of Stanford University and is currently in her first year of a creative writing master's program at the University of Texas. Her story, "Gypsy Quiet," was published in the Spring 2005 issue of *Appalachian Heritage.*

Stephen George is a staff writer for the *Louisville Eccentric Observer* (LEO). Born and raised in Louisville, George is a relatively content resident of the Germantown neighborhood. His experience on the KFTC mountaintop-removal tour, coupled with the bits of childhood he spent with relatives in the bright and wide fields of Lebanon, Kentucky, has considerably affected his world view.

Jerry Hardt is the communications/fundraising director for Kentuckians For The Commonwealth, and he serves as editor of the newspaper *balancing the scales.* He grew up in Louisville and lives in Magoffin County with wife Diane and daughter Elizabeth. He was instrumental in the creation of *Missing Mountains.*

Chris Holbrook, a Knott County native, is the author of *Hell and Ohio: Stories of Southern Appalachia* (Gnomon Press, 1995). A graduate of the University of Kentucky, he has won fellowships from the Fine Arts Work Center in Provincetown and the Kentucky Arts Council. His short stories have won competitions sponsored by *Now and Then* and *Louisville Magazine.* An assistant professor of creative writing at Morehead State University, he lives in Lexington with his wife, Mary Beth, and his daughter, Erin.

Silas House is the author of the novels *Clay's Quilt, A Parchment of Leaves,* and *The Coal Tattoo,* as well as the play, *The Hurting Part.* He has received the James Still Award for Special Achievement from the Fellowship of Southern Writers, the Chaffin Award for Literature, the Kentucky Novel of the Year Award, and two ForeWord Magazine Bronze Awards for Best Literary Novel. He is the Writer-In-Residence at Lincoln Memorial University.

Charlie Hughes was raised on a farm bordered by the Salt River near McAfee, Kentucky. He often revisits that rural locale in both his fiction and poetry. He is the former editor of *Wind* Magazine and author of *Shifting for Myself,* a collection of poems. He is also owner of Wind Publications, a small press, and editor of *The Kentucky Literary Newsletter.*

Kristin Johannsen is an environmental and travel writer in Berea. Her articles have appeared in *The Los Angeles Times*, *The Chicago Tribune*, *Mother Jones*, and other publications. Her latest book is *Ginseng Dreams: the Secret World of America's Most Valuable Plant* (University Press of Kentucky, 2006.) She is one of the editors of *Missing Mountains*.

Leatha Kendrick is the author of two volumes of poetry: *Heart Cake* (Sow's Ear Press) and *Science in Your Own Backyard* (Larkspur Press). She also wrote the screenplay for the documentary film *A Lasting Thing for the World*. She has taught creative writing at the University of Kentucky and elsewhere. She recently co-edited (along with George Ella Lyon), *Crossing Troublesome, Twenty-five Years of the Appalachian Writers Workshop* (Wind, 2002).

Loyal Jones has spent almost all of his life working with Appalachian people, first as associate director and director of the Council of the Southern Mountains and later as founder and director of the Berea College Appalachian Center. His latest book is *Faith and Meaning in the Southern Uplands*. He lives in Berea.

Christina Lovin's poems have appeared in *The Harvard Summer Review*, *The Northern New England Review*, *Hunger Mountain*, *Entelechy International*, *Off the Coast*, and other periodicals and anthologies. She is a recipient of both a 2005 Al Smith Professional Assistance Grant and a 2005 Professional Development Grant from the Kentucky Arts Council. A resident of Lancaster, Kentucky, she teaches college writing classes in and around Lexington.

George Ella Lyon, originally from Harlan County, has worked with the Appalachian Poetry Project, the Grassroots Poetry Project, and the Appalachian Writers Workshop. She has co-edited three collections of writing from the region. Her most recent books are *Weaving the Rainbow* and *Sonny's House of Spies*. She makes her living as a freelance writer and teacher in Lexington.

Maurice Manning, winner of the Yale Younger Poets Award, is the author of the poetry collections *Lawrence Booth's Book of Visions* and *A Companion for Owls; Being the Commonplace Book of D. Boone, Long Hunter, Back Woodsman, &c.* A native of Danville, Kentucky, he currently teaches creative writing at Indiana University—when he is not homesteading at his farm in Mercer County.

Ed McClanahan grew up in northeastern Kentucky, in Brooksville and Maysville. His fiction and non-fiction has appeared in *Esquire*, *Playboy*, *Rolling Stone*, and many other magazines. He is also the author of *The Natural Man*, *Famous people I Have Known*, and three other books. He edited *Spit in the Ocean#7: All About Kesey* (Viking-Penguin, 2003), a tribute to the late Ken Kesey. Ed and his wife, Hilda, live in Lexington with their Great Dane, Frieda.

Davis McCombs grew up in Hart County, Kentucky. His first book, *Ultima Thule*, was chosen as the winner of the 1999 Yale Series of Younger Poets. "Tobacco Mosaic," a sequence of 16 poems about burley-tobacco farming, appears in the Summer 2005 issue of *The Missouri Review* as the winner of the Larry Levis

Editor's Prize. New work has also recently appeared, or is forthcoming, in *Poetry, The Kenyon Review, Willow Springs,* and *Pleiades.* Davis teaches in the MFA Program at the University of Arkansas.

Bobbie Ann Mason is the award-winning author of numerous books, including *In Country, Shiloh and Other Stories, Feather Crowns,* and *Clear Springs.* Born and raised in Mayfield, she now lives in Anderson County. She is Writer-in-Residence at the University of Kentucky. Her latest novel is *An Atomic Romance* (Random House, 2005). Her next work of fiction, *Nancy Culpepper,* is due out in 2006. She is an editor of *Missing Mountains.*

Amanda Moore is a native Kentuckian and former staff attorney with the Appalachian Citizens Law Center in Prestonsburg. She currently lives in Morehead with her husband and infant son. Her essay, "The Law Leaves a Hollow Place," was written especially for *Missing Mountains.*

Daymon Morgan was born on Camp Creek in Leslie County. He has always hunted and dug herbs there and across the mountain on Lower Bad Creek, where he bought a place in 1947. He still lives there, despite problems caused by coal operators. He is a past chairperson of Kentuckians For The Commonwealth.

Gurney Norman, former head of the creative writing program at the University of Kentucky, is the author of the widely praised novel *Divine Right's Trip: a Novel of the Counterculture* (which first appeared in *The Last Whole Earth Catalog*) and the short story collection *Kinfolks: the Wilgus Stories.* Raised near Hazard, Kentucky, he now lives in Lexington with his wife, Nyoka.

Ann W. Olson, who has lived on Mauk Ridge in Elliott County in northeastern Kentucky for over thirty years, became a nature photographer and children's book illustrator eight years ago. She photo-illustrated *Counting on the Woods,* written by George Ella Lyon.

Erik Reece's book, *Lost Mountain: Radical Strip Mining and the Devastation of Appalachia,* will be published by Riverhead in February, 2006. His work has appeared in *Harper's, Orion, The Oxford American,* and other places. He lives in Lexington with his wife Mary.

Susan Starr Richards has lived in Kentucky for forty years. She and her husband raised thoroughbreds on their farms in Harrison and Scott Counties. Larkspur Press has just published her book of poems (*The Life Horse,* 2005), and in 2006, Sarabande Books will publish her story collection, *The Hanging in the Foaling Barn,* as the third volume in the Woodford Reserve Series in Kentucky Literature.

Gwyn Hyman Rubio is the author of the novels *Icy Sparks* (a selection of Oprah's Book Club) and *The Woodsman's Daughter.* Her short story, "Little Saint," won a First Prize Cecil Hackney Literary Award. Daughter of Mac H. Hyman, the author of *No Time for Sergeants,* she grew up in Cordele, Georgia. A former Peace Corps and Vista volunteer, she now lives with her husband in Versailles, Kentucky.

Anne Shelby, a native of southeastern Kentucky, lives near Oneida in Clay County. Her columns appear regularly in *The Lexington Herald-Leader* and other Kentucky newspapers. Forthcoming publications include *The Man who Lived in a Hollow Tree*, a picture book from Simon & Schuster, and *Appalachian Studies*, a poetry collection from Wind Publications.

Bob Sloan is the author of a short story collection, *Bearskin to Holly Fork*, and *Home Call: A Novel of Kentucky*. His novel *Nobody Knows, Nobody Sees* is forthcoming in the spring of 2006. Bob and his wife Julie live east of Morehead, on a small farm that belonged to his grandfather and his father.

Pauline Stacy was born in the Bluegrass Coal Camp in what is now part of Hazard. Looking for a smaller community, in 1973 she and her husband moved to the Ary, a community in eastern Perry County, where her husband was born and raised. They have two children and two grandchildren.

Richard Taylor, a former Kentucky Poet Laureate, teaches at Kentucky State University, and with his wife Lizz owns Poor Richard's Bookstore in Frankfort. His most recent work is *Stone Eye*, a volume of poetry from Larkspur Press (2004). His novel *Sue Mundy* is forthcoming from the University of Kentucky Press.

Mary Ann Taylor-Hall, whose fiction has appeared in *Best American Short Stories*, is the author of the acclaimed novel *Come and Go, Molly Snow,* and the short story collection *How She Knows What She Knows about Yo-yos*, which was named the best book of short fiction in 2000 by *Foreword Magazine*. She is an editor for *Missing Mountains*. She has lived on the Harrison-Scott County line for thirty years.

Betty Woods grew up in the Leatherwood coal camp and moved to the Ary community in Perry County in the mid-1990s. Her personal statement in *Missing Mountains* is based on an interview she gave to Kentuckians For The Commonwealth.

Jeff Worley, who came to Kentucky in 1986, has a new chapbook of poetry, titled *Leave Time*, from Finishing Line Press in Georgetown, KY. His third book-length collection, *Happy Hour at the Two Keys Tavern*, will be published by Mid-List Press in April 2006. His poems have appeared widely, in such magazines as *Poetry Northwest, The Georgia Review, The Atlanta Review, New England Review, Shenandoah, The Southern Review, and The Sewanee Review*.

ACKNOWLEDGMENTS

"Forced from Home" originally appeared in *balancing the scales*, May 23, 2004. Reprinted by permission of Kentuckians For The Commonwealth

.

"Statement on Mountaintop Removal," by 16 Kentucky Writers, "Getting Quiet," by Whitney Baker, "Mining Destroys Land and Spirit," by Silas House, "Letter to the Editor," by Ed McClanahan, "The Facts Aren't Pretty," by Erik Reece, "For Love of Kentucky, Stop Destroying Her," by Bob Sloan, "The Coal Industry's False Fronts," by Bob Sloan, "Truth Buried Under Natural Materials," by Anne Shelby, and "The Ghosts of Mountains," by Mary Ann Taylor-Hall, all appeared originally in *The Lexington Herald-Leader* (sometimes in slightly different form and/or under different titles). Reprinted by permission of the authors.

"Contempt for Small Places" and "Compromise, Hell!" copyright 2005 by Wendell Berry, from *The Way of Ignorance*. Reprinted by permission of Shoemaker & Hoard Publishers (an Avalon Publishing Group imprint).

"Bringing Down a Mountain" by Stephen George originally appeared in LEO, May 18, 2005. Reprinted by permission of LEO.

"Visiting the Site of One of the First Churches my Grandfather Pastored" by Tony Crunk originally appeared in *Living in the Resurrection*. Reprinted by permission of Yale University Press.

"Upheaval," by Chris Holbrook, originally appeared in *Nighttrain*. Reprinted by permission of the author.

"Stripped," by George Ella Lyon, first appeared in *The Appalachian Journal* and was reprinted in *Where I'm From, Where Poems Come From*, Absey & Company, 1999.

"The Sum Result of Speculation" copyright 2004 by Maurice Manning, reprinted from *A Companion for Owls, Being the Commonplace Book of D. Boone Long Hunter, Back Woodsman, &c.*, published by Harcourt, Inc., 2004.

"Big Bertha Stories," appeared in *Love Life*, HarperCollins, reprinted by permission of International Creative Management, Inc. Copyright ©1989 by Bobbie Ann Mason.

Betty Woods' words are excerpted from an April 1999 interview published by Kentuckians For The Commonwealth.